T0156003

Mechanism of Fires

Ali S. Rangwala · Vasudevan Raghavan

Mechanism of Fires

Chemistry and Physical Aspects

Ali S. Rangwala
Department of Fire Protection Engineering
Worcester Polytechnic Institute
Worcester, MA, USA

Vasudevan Raghavan
Department of Mechanical Engineering
Indian Institute of Technology Madras
Chennai, Tamil Nadu, India

ISBN 978-3-030-75500-3 ISBN 978-3-030-75498-3 (eBook)
https://doi.org/10.1007/978-3-030-75498-3

Jointly published with ANE Books India
In addition to this printed edition, there is a local printed edition of this work available via Ane Books in South Asia (India, Pakistan, Sri Lanka, Bangladesh, Nepal and Bhutan) and Africa (all countries in the African subcontinent).
ISBN of the Co-Publisher's edition: 978-9-389-21278-5

This Springer imprint is published by the registered company Springer Nature Switzerland AG
The registered company address is: Gewerbestrasse 11, 6330 Cham, Switzerland

Preface

We are happy to bring out this book on Mechanism of Fires—Chemistry and Physical Aspects. This book is intended for post-graduate students in Mechanical, Chemical, Fire protection, and allied engineering disciplines. Further, this book will be helpful to scientists and engineers working in the areas of fire protection to recapitulate the fundamental and applied aspects of fire.

The aim of the book is to make the readers swiftly understand necessary aspects of fires with sufficient details, make them capable of identifying and referring additional reference books and manuscripts published in fire related journals as per the requirement, and begin the research on fire studies.

The unique features of the book are (1) a comprehensive description of fuels involved in fires, definitions related to fire, thermodynamics for fire calculations, basics of transport processes and fundamental aspects of combustion related to fire, (2) physical descriptions of premixed and nonpremixed flames, (3) a detailed analysis of the characteristics of fires from solid and liquid fuels, including ignition, spread, and burning rates, and (4) physical aspects of fire plumes, compartment fires, and dust fires. The unique features of the book are (1) a comprehensive description of fuels involved in fires, definitions related to fire, thermodynamics for fire calculations, basics of transport processes and fundamental aspects of combustion related to fire, (2) physical descriptions of premixed and nonpremixed flames, (3) a detailed analysis of the characteristics of fires from solid and liquid fuels, including ignition, spread and burning rates, and (4) physical aspects of fire plumes, compartment fires, and dust fires.

The fundamental aspects related to evaporation of liquid fuels and pyrolysis of solid fuels are explained with simplified mathematical expressions. Worked examples are provided as required to illustrate mathematical calculations involved in fire analysis. Case studies presented in the chapters are obtained from carefully done lab-scale experiments as well as from comprehensive numerical simulations.

Review questions provided at the end of the chapters help the readers to evaluate their understanding of both the fundamental as well as the application aspects. Apart from standard text and reference books, several important research articles have been cited in the text and listed in the references.

Worcester, USA Ali S. Rangwala
Chennai, India Vasudevan Raghavan

Acknowledgements

We wish to acknowledge the contributions made by Dr. Sharanya Nair in typesetting the book chapters and for preparing high quality figures where necessary. She has also contributed to reviewing through the chapters and assisting in preparing review questions and case studies. We also wish to acknowledge Dr. S. Muthu Kumaran for his contributions to numerical simulations, numerical problems and for proof-reading the chapters.

We also wish to acknowledge the many researchers who have worked in the Fire Protection Engineering Laboratory of WPI and have contributed to the contents presented in this book. Mohammad Alshuqaiq, Shivaprasad Arava, Kemal Arsava, Yann Bouchaud, Peter Bellino, Trevor Borth, Amy Brown, Jay Brown, Kelley Brown, Li Chang, Jerimiah Crocker, Julia Cuendet, Giovani Di Cristina, Jason Daniels, Andre Da Vitoria, Brian Elias, Hamed Farahani, Cécilia Florit, Yanyun Fu, Laurens vanGelderen, Michael Gollner, Barbara Hall, Kathryn Hall, Todd Hetrick, Terumi Hisano, Hsin-Hsiu (Matt) Ho, Kulbhushan Joshi, Furkan Kodakoglu, Veronica Kimmerly, Mahesh Kottalgi, Shijin Kozhumal, Suhas Lakkundi, Minkyu Lee, Jonathan Levin, Cong Li, Haoran Li, Glenn Mahnken, Nadia Mofidi, Kristopher Overholt, Haejun Park, Dave Petrow, Xiaoyue Pi, Raymond Ranellone, Sreenivasan Ranganathan, Scott Rockwell, Marine Roux, Samim Safaei, Nate Sauer, Hayri Sezer, Lydia Shi, Joel Sipe, Panyawat Tukaew, Orlando Ugarte, Cédric Venet, ApoorvWalawalkar, Yu Wang, Chuming Wei, Yanxuan Xie, Ying Zou. Several discussions and problems in the book have resulted from M.S. and Ph.D. theses originating in the FPE-WPI. Rangwala would also like to especially thank Prof. Jose Torero for useful discussions and guidance during the preparation of the book.

Ali S. Rangwala
Vasudevan Raghavan

Contents

About the Authors

Dr. Ali S. Rangwala is a Professor of Fire Protection Engineering at Worcester Polytechnic Institute (WPI). He has a Ph.D. in Mechanical and Aerospace Engineering from the University of California, San Diego. He received the NSF CAREER award on Controlling Mechanisms of Dust Layer Ignition and Flame Propagation in Dust Clouds. He has also served on panels for the National Science Foundation (NSF), Occupational Safety and Health Administration (OSHA), National Fire Protection Association, (NFPA), and National Aeronautics and Space Administration (NASA), related to areas of combustion, fire and explosion. He has 4 patents related to different topics in combustion. He has published over 60 peer-reviewed international journal articles and has presented in over 70 conferences. He teaches three graduate courses: Explosion Protection, Industrial Fire Protection, and Combustion at the Department of Fire Protection Engineering in WPI.

Dr. Vasudevan Raghavan is currently Professor in the Department of Mechanical Engineering, Indian Institute of Technology Madras (IIT), India. He obtained his Ph.D. from IIT Madras and has carried out his postdoctoral research in the University of Nebraska-Lincoln, USA, in NASA aided microgravity droplet combustion simulations. His areas of research include computational fluid dynamics applied to reacting flows, laminar flames, fire modelling, evaporation and combustion of liquid fuel droplets, flame spread and liquid fuel pool combustion, gasification and combustion of coal and biomass. He has authored about 110 international peer-reviewed journal articles and 60 international conference articles. He teaches graduate courses such as Theory of Fire Propagation, Fundamentals of Combustion, Combustion Technology and Applied Thermodynamics at the Department of Mechanical Engineering in IITM.

Chapter 1
Basic Considerations

Fire is a rapid exothermic oxidation reaction (combustion reaction) involving several types of fuels. Heat and toxic combustion products are released during a fire. It is generally turbulent in nature and predominantly consists of gas-phase reactions. Fire is a combustion process that occurs usually in non-premixed mode (as diffusion flames). The heat generated from the fire helps in gasifying solid and liquid fuels, while air from the ambient provides necessary oxygen for the oxidation reaction. These two reactants feed the fire (reaction zone) to continue the cycle. Hence the heat release, the oxygen entrainment, and the fuel availability are coupled, forming the main elements for the fire. The fire ceases to exist once one of these elements depletes. The interaction of these elements through mixing of reactants and heat and mass transfer processes is enhanced by the induced flow field around the fire. The temperature variations in the fire zone create density differences, which alter the flow field due to the buoyancy force. Thus, in several fire scenarios, natural convection of wide range of scales exists:

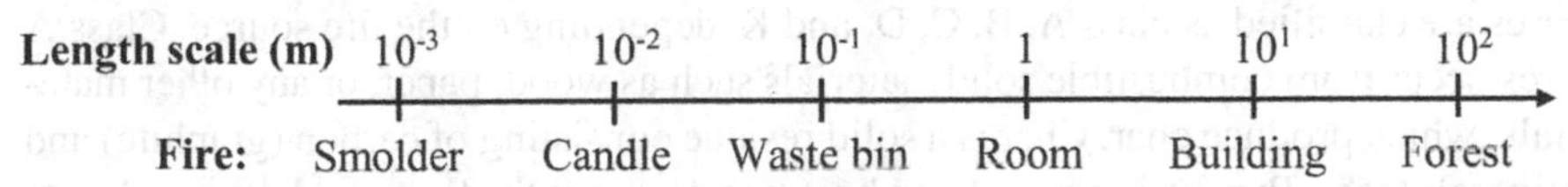

In heterogeneous fire scenario, condensed phase reactions are also observed over the surfaces of solid fuels. For instance, wood typically consists of moisture, gaseous components called volatiles, fixed carbon, and minerals called ash. When heated or partly burnt, moisture and volatiles leave the wood, and carbon and ash remain. If this residue (char) is hot enough, the ambient air that diffuses towards its surface reacts

A. S. Rangwala and V. Raghavan, *Mechanism of Fires*,
https://doi.org/10.1007/978-3-030-75498-3_1

with it. No flame is seen; however, the surface becomes incandescent (glowing with bright red color). This is called smoldering process. Smoldering occurs either before or after the occurrence of fire (flame mode). It results in the release of hot products (in the form of smoke) as well as some fuel species.

Fire can be caused by natural means such as lightning, volcanoes, earth quakes as well as by underground fires, usually caused by self-heating of coal. These fires are often huge. For example, the Ruth Mullins underground coal fire in Kentucky, USA has been burning for more than 4 decades and spans a vast network of underground smoldering regions spanning several square miles. Similar underground coal fires are found in several regions of the world [1]. Another example usually observed in urban and forest fires is the formation of a fire whirl, which displays a high intensity burning at a rapid rate, such as the Kanto fire whirl in Japan that resulted in the death of 38000 people in 15 min! [2]. Fire can also be human-made like in wars or by accidents like in domestic and industrial fires.

Several materials are involved in fires. Wood (cellulose), polymers, and fabrics are examples of common sources in domestic fires. Solvents, gases, and flammable chemicals are examples of fire sources in industries, while wild (forest) fires are caused by natural vegetation.

In this chapter, basic terminologies, classification, thermal and physical properties of materials, and non-dimensional numbers associated with fires are discussed. The physical and chemical aspects of fires should be understood to prevent the occurrence of fire as well as to extinguish the fire without causing loss of life, property, and structure, in case of a fire accident. A good knowledge of fire sources—flammability or combustible characteristics of the materials and their properties—is important to make proper usage of these materials in several applications.

1.1 Classification of Fire Sources

Fires are classified as class A, B, C, D, and K, depending on the fire source. Class A fires occur from combustible solid materials such as wood, paper, or any other materials, which produce char. Char is a solid residue consisting of carbon (graphite) and minerals (ash). Pyrolysis occurs upon heat transfer from the flame and the pyrolysate (released volatile gases) participate in the combustion reaction. These gas-phase reactions sustain over the solid material. When all the volatiles are released from the solid, char is left over. Char is further heated up by heat transfer from the flame. When oxygen penetrates towards the hot char surface, char oxidation occurs in the form of surface reaction. An incandescent (red hot) char layer is visible during the surface oxidation process. This type of reaction is also termed as smoldering. These fires are quenched by cooling the fuel to stop the pyrolysis.

Liquid and gaseous fuels contribute to class B fires. Evaporation of liquid fuels is much simpler when compared to the pyrolysis of solid fuels. When a liquid pool receives heat from the hot flame through its surface, a thermodynamic equilibrium prevails at the interface. Based on the equilibrium interface temperature, a definite mass fraction of fuel exists at the gas phase close to the interface. The fuel mass from the interface is transported into the gas phase both by convection and diffusion (predominantly due to concentration gradient) as governed by Fick's law. This fuel vapor mixes with oxygen from the ambient and makes the non-premixed flame to sustain over the pool surface. The rate of burning of liquid pool depends upon the volatility of the liquid fuel, ambient conditions, and gas-phase convection. Non-charring solids, which melt and vaporize or sublimate (vaporize from solid state) during burning, display characteristics of liquid fuels. Agents such as foams are used to quench liquid pool fires in copper mines, which use kerosene pools as solvents. Fires from gaseous fuels are generally hazardous as additional rate limiting process such as gasification or evaporation is not present during their burning. These fires can be quenched by diluting the oxygen, reducing the temperature or by inhibiting the chemical reaction. Agents that inhibit the chemical reaction process are effectively used to quench fires from gas fuels.

In class C fires, energized electrical devices, such as electronic devices, computers, motors, and transformers, are involved. Apart from fire hazards, electric shock and reignition of fire as a result of electrical spark may occur. The disconnection of power is the first step in the process of inhibition of such fires. Class D fires occur from combustible metals. Here, the heat release is quite intensive, with the flame temperature being more than 2000 °C. For class C and class D fires, special extinguishers are used. For instance, water cannot be used in these cases, especially if power cannot be disconnected in class C fires and in the case of class D fires, metals like sodium and lithium react with water in an exothermic manner. Dry sand or powders, which are non-reactive with the metals, and non-conductive, are used to quench such fires.

Class K classifies fires from cooking oil, greases, animal, and vegetable fats. Fires of this category occur as a result of heated up material well above its fire point. Wet chemical agents such as potassium acetate or potassium carbonate are used to extinguish such fires.

1.2 Properties of Combustible Materials

Calorific value (J/kg) is one of the important properties for all the combustible materials. It is the heat released when one kg of the material is completely burnt. Based on the state of the material, whether it is solid, liquid, or gas, several other properties are defined. These properties contribute to the analysis of processes such as pyrolysis, ignition, flame propagation, and rate of burning of these materials. Important thermo-physical properties for all types of combustible materials include *thermal conductivity*, k, (W/m-K), *specific heat*, C, (J/kg-K) and *density*, ρ (kg/m^3). A useful property called *thermal diffusivity* (m^2/s), is defined in terms of these three properties, as $k/(\rho \cdot C)$. The product of these three properties ($\rho \cdot k \cdot C$), having units of [{W/(m^2·K)}^{2}·s], is called *thermal inertia*.

Stoichiometric air is the quantity of air (kg) required to burn one kg of the material completely. *Standard heat of combustion* (J/kg-fuel) is the heat released when one kg of combustible material from standard condition (1 atm, 298 K) is burnt completely using air and the products are cooled to 298 K. These two are important quantities used in fire analysis. The standard heat of combustion is termed as calorific value. It is called lower calorific value when all the water in the product of combustion is in vapor phase and called higher calorific value if water is in liquid phase. This quantity is used to determine the heat release rate using the rate at which the material is burnt. Heat of combustion is also written in terms of per kg of oxygen (J/kg-O_2) and is calculated based on the chemical composition of the fuel. Air, under standard conditions, is assumed to be composed of 21% O_2 and 79% N_2 by volume.

In order to calculate the calorific value, standard enthalpies of the constituents of fuel and air are required. *Standard enthalpy* is a specific property (J/kmol), defined as the sum of standard *enthalpy of formation* (enthalpy associated with the formation of the material from its naturally occurring elements) and *sensible enthalpy* (enthalpy rise due to increase in temperature from 298 K to the given value). Another important quantity used in fire analysis is the *adiabatic flame temperature*, which is obtained when there is no loss of heat from the fire to the ambient (theoretical quantity). It is the temperature of the products obtained from complete combustion of the fuel material, at which the enthalpy of the reactants equals the enthalpy of the products.

For liquid fuels, three temperatures are defined in increasing order; *flash point*, *fire point*, and *boiling point*. Flash point is the temperature at which enough vapor is released from fuel surface to form a mixture with atmospheric air that is ignitable. Below this temperature, even though vapors are formed over the interface, they are not sufficient to form an ignitable mixture. At flash point, ignition is possible only with the help of a pilot ignition source such as spark or small flame. At a still higher temperature, called fire point, vapors are readily available to feed a non-premixed flame. At the flame zone, fuel vapor and oxygen from air mix around stoichiometric proportions to sustain the flame without the need of pilot ignition. Boiling point is the saturation temperature, where phase change occurs without a change in the temperature. When a liquid pool burns by having heat transfer only through its surface, its surface temperature will not exceed its boiling point. The heat transfer from the gas phase to interface is partly used to supply the energy for phase change, called the *latent heat of vaporization* (J/kg) and rest of it is supplied for heating the liquid phase.

One more notable property of liquid fuel is its *surface tension* (N/m). The gradients of surface tension along the interface trigger flow at the liquid surface. This is called Marangoni convection.

Solid fuels are classified as *charring* and *non-charring* fuels. Examples of charring fuels are wood and coal. As mentioned earlier for the case of wood, charring fuels have four basic constituents, namely, *moisture*, *volatiles*, *fixed carbon*, and inert mineral material called *ash*. When heat is transferred to the surface of a charring solid, moisture is released first around a temperature close to 373 K. Upon further heating to a higher temperature (~500 K to 600 K), volatiles are released. Remaining fixed carbon and ash are together called *char*. The fixed carbon content of the char, when hot enough, also burns at a slower rate as a result of oxygen diffusion towards its surface. This is a *heterogeneous reaction* occurring at the char surface, as opposed to *homogeneous reactions* occurring at the gas phase. Char combustion onsets at its *ignition temperature*. Finally, ash is left behind as a residue.

On the other hand, non-charring solid fuels either melt to liquid fuels or monomers, or directly sublimate to vapor. Wax and polymers are examples of non-charring fuels. Only impurities, if any, are left behind after combustion. Pyrolysis is the process in which gaseous components come out of the solid fuel. The major heat release reactions occur in gas phase and the gaseous components leaving the solid fuel to participate in these gas-phase reactions. The amount of heat supplied to the solid fuel to extract the gaseous components from it is called *heat of pyrolysis* (J/kg) or *heat of gasification*. For sublimating non-charring solids, *latent heat of sublimation* (J/kg) is supplied at a given temperature to convert the solid to gas. To melt a solid, *latent heat of melting* (J/kg) is supplied at the solid's melting point.

It should be kept in mind that most of these properties vary notably with temperature. Empirical correlations are often used to determine the property at an average temperature to be used in fire calculations. In numerical models, all the properties are evaluated as a function of the temperature at a given location. Since gas phase is constituted by multiple species, a proper mixing rule is also used to evaluate the properties of the mixture.

1.3 Definitions Related to Fire

Several important parameters and definitions are used in fire analysis. These are discussed in this section. For a material to be ignited, its surface should reach a minimum temperature, called *ignition temperature*. To attain this, the surface should be subjected to an adequate heat flux. A part of the heat incident on the surface is conducted into the material, depending upon the thermal inertia of the material. Some amount of heat is lost to the surroundings by radiation and is called *surface re-radiation loss*. The remaining amount is used to generate gases from the condensed surface. The minimum heat flux at which ignition occurs is called *critical heat flux*. The corresponding mass of fuel vapor released per unit area is called *critical mass burning rate*. The time delay for ignition to occur is related to *thermal response parameter*. This parameter depends on thermal inertia, heat of gasification, ignition temperature, and *product yield* during pyrolysis.

Ignition is classified as *piloted ignition* and *autoignition*. The heat source, which initiates the pyrolysis of a material, usually contributes to the ignition energy that the pyrolysate and air mixture requires for ignition. The ignition caused by an external energy source is called piloted ignition. The requirements for piloted ignition are formation of required quantity of the flammable reactant mixture (*critical volume*) and *minimum ignition energy* that overcomes the heat loss. For a reactant mixture to be flammable, it should be within its *lower flammability* and *upper flammability* limits. In the former, a fuel lean reactant mixture with minimum amount of fuel exists that allows the flame to propagate through it. In the latter, a fuel-rich reactant mixture with minimum amount of oxidizer is present, which allows the flame to propagate through it.

Autoignition occurs if the reactant mixture is present within the flammability limits and is also at a high temperature, called *autoignition temperature*. This condition allows the reactant mixture to ignite without the need of an external heat source. The phenomenon of autoignition also occurs in a layer of combustible dust settled over a hot surface. Here, based on the thickness of the dust layer and the temperature of the hot surface, a temperature rise beyond the surface temperature is observed. This is referred to as *spontaneous ignition*. Autoignition of reactant mixtures present in confined space may lead to an *explosion*. Here, heat release causes an increase in both temperature as well as pressure. This scenario is also observed in compartment (room) fires, where the ventilation is much less. Simultaneous autoignition of several commodities in a compartment is referred to as *flashover*. Autoignition also causes a phenomenon called *backdraft*. This happens when oxygen starved hot fuel vapors are suddenly exposed to a stream of air from the atmosphere.

Once a flame is initiated over the fuel surface, a parameter called *flame heat flux* dictates its sustained burning. Flame heat flux is the fraction of heat transferred from the flame back to the surface. The heat released from the flame per unit mass of the material pyrolyzed is called the *heat of combustion*. Part of this heat is transferred by the product gases during their transport and a part is transferred from the flame (hot zone) by radiation. The former is referred to as *convective heat flux* and the latter is called *radiative heat flux*, calculated per unit area of the fire source. The surface receives the remaining heat by conduction, convection as well as radiation (depending upon the absorptivity of the surface). The ratio of the heat released in fire from unit mass of material pyrolyzed to the heat required to pyrolyze unit mass of the material is called *heat release parameter*.

After the onset of ignition, the extent and the rate of spread of a flame relative to the region of ignition is quantified by *flame propagation index* (FPI). Higher the value of FPI, higher will be the flame propagation rate over the material. Ignition and fire propagation (spread) also depend upon geometrical parameters such as material thickness. Based on this, a solid material is classified as *thermally thick* or *thermally thin*, depending upon the temperature gradient that would set in within the solid, when the solid receives heat flux from the ambient. Flame spread in solid is also affected by the *orientation* of the material with respect to the gravity vector. Gas-phase convection also dictates the rate of flame spread. This includes both natural convection and forced convection, as shown in Fig. 1.1. If the convective flow is in the same direction as the direction of flame spread, then it is termed as *concurrent flame spread*. On the other hand, if the flame spreads in the opposite direction to the gas-phase convective flow, it is called *opposed flame spread*. These are illustrated for both natural and forced convection scenarios in Fig. 1.1. Concurrent flame spread is rapid and unsteady. Opposed flame spread is gradual and steady. After the flame spreads over the entire surface of the material, depending upon the material thickness, the mass is lost at almost a steady rate. This is called *mass burning rate*. Here, the heat transferred from the fire to the fuel is primarily used to retrieve the fuel gases

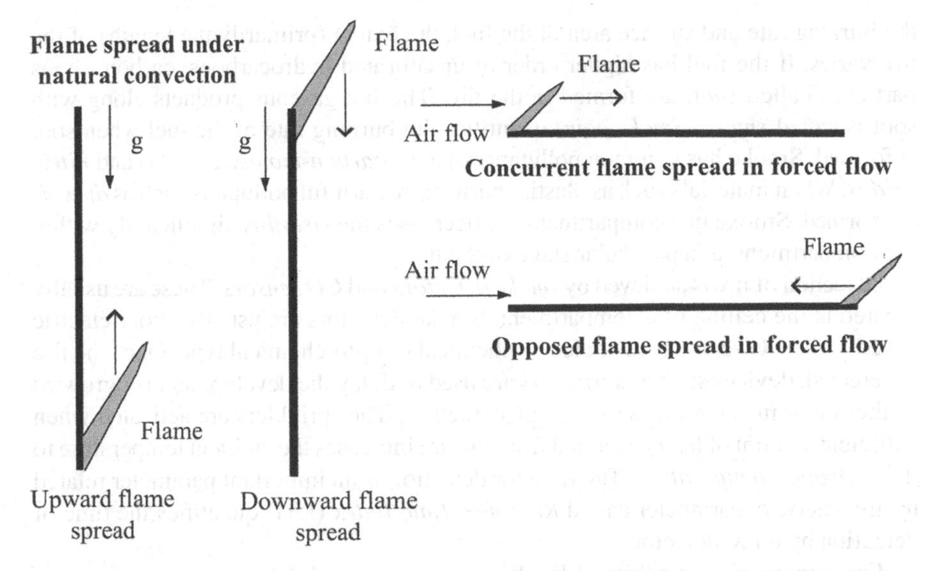

Fig. 1.1 Flame spread modes over fuel surfaces under natural and forced convection

Fig. 1.2 Steady burning over fuel surfaces under natural convection

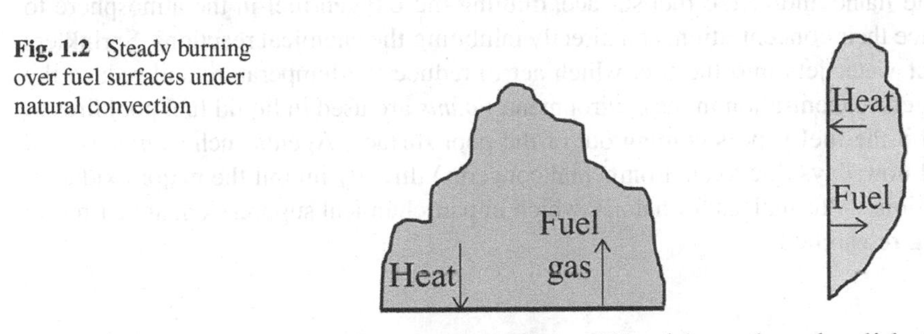

from the surface of the condensed fuel. This is illustrated in Fig. 1.2. Mass burning rate is dictated by the fuel thickness, gas-phase convection, and heat loss, apart from the material properties.

As mentioned before, charring materials undergo heterogeneous surface reaction, by which oxidation of char takes place. This reaction is different from the *flaming reaction* occurring in the gas phase. Here, no flame is observed and a red colored hot zone is seen as in the case of cigarette burning. This is called *smoldering* combustion. Porosity of the material, oxygen concentration and its diffusion, and heat loss to the ambient dictate the smoldering process. Smoldering can initiate a flaming combustion (fire) or at the end of the flaming combustion, smoldering can occur.

Smoldering occurring in oxygen limited environment inside a compartment leads to generation of hot fuel gases, which can auto-ignite when exposed to sufficient amount of atmospheric air. Since fire is a turbulent non-premixed flame, based on the burning rate and surface area of the fuel, the extent (primarily the length) of the fire varies. If the fuel has higher order of unsaturated hydrocarbons, carbonaceous particles, called *soot*, are formed in the fire. The hot gaseous products along with soot is called *smoke*. *Smoke point* quantifies the burning rate of the fuel when soot is formed. Smoke has common pollutants such as *carbon-monoxide* (CO) and *nitric oxides*. When materials such as plastics burn, more harmful pollutants such as *dioxins* are formed. Smoke in a compartment fire decreases the *visibility* significantly within the compartment, at a particular stage of the fire.

Detection of fire is achieved by *smoke detectors* and *CO sensors*. These are usually located in the ceiling of a compartment. Smoke detectors are usually photo-electric based while CO sensors are of electro-chemical or opto-chemical type. Once the fire is detected, devices such as *sprinklers* are used to delay the development (or growth) of the fire or in some cases to extinguish the fire. The sprinklers are activated when sufficient amount of heat generated from the fire increases the ambient temperature to the *activation temperature*. The time for detection is an important parameter related to fire safety. A parameter called *Response Time Index* (RTI) quantifies the time of detection by a fire detector.

Fire suppression is achieved by three primary ways; reducing the temperature of the flame and/or the fuel surface, diluting the oxygen/fuel in the atmosphere to reduce their concentration, and directly inhibiting the chemical reactions. Sprinklers inject water jets into the fire, which act to reduce the temperature and reduce the oxygen concentration in the environment. *Foams* are used in liquid fuel pool fires to inhibit the fuel vapors coming out of the pool surface. Agents such as *halons* (not used nowadays due to environmental concerns) directly inhibit the major oxidation reactions. Alternatives for halons, which impart chemical suppression, are currently being researched.

1.4 Flammability Tests

Flammability tests are carried out to evaluate several fire related aspects of materials. These include tests for ignitability, flame spread rate, mass burning rate, and heat release rate. ASTM provides standards to carry out these tests. The webpage link of the ASTM standards is given in Ref. [3]. The apparatus commonly used for flammability tests are *flame propagation apparatus* (FPA) and *cone calorimeter* (CC). Schematics of typical CC and FPA are shown in Figs. 1.3 and 1.4, respectively. Schematic of an intermediate scale FPA, analyzing burning of vertically oriented sample is shown in Fig. 1.5.

Ignition tests are conducted on materials by subjecting its surface to a certain heat flux and using a pilot igniter, such as a spark plug, at a given distance from the material's surface. Heat flux at which flaming ignition occurs is called critical heat flux. Temperature of the surface, at which ignition occurs, is called ignition temperature. Thermal response parameter (TRP) is then defined using thermal inertia of the material and ignition temperature. Empirical correlation for time for ignition

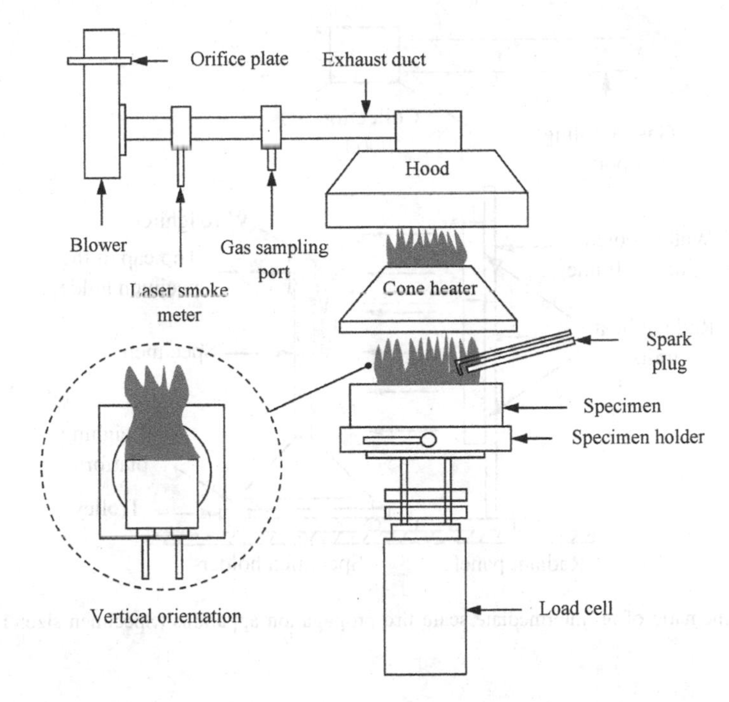

Fig. 1.3 Schematic of cone calorimeter (Specimen size: 0.1 m × 0.1 m) [4]

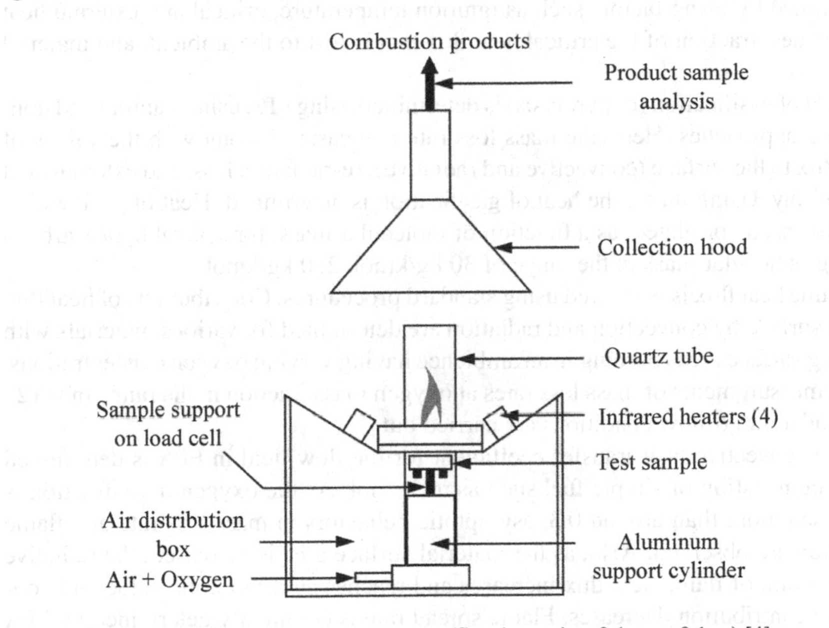

Fig. 1.4 Schematic of fire propagation apparatus (Specimen size: 0.1 m × 0.1 m) [4]

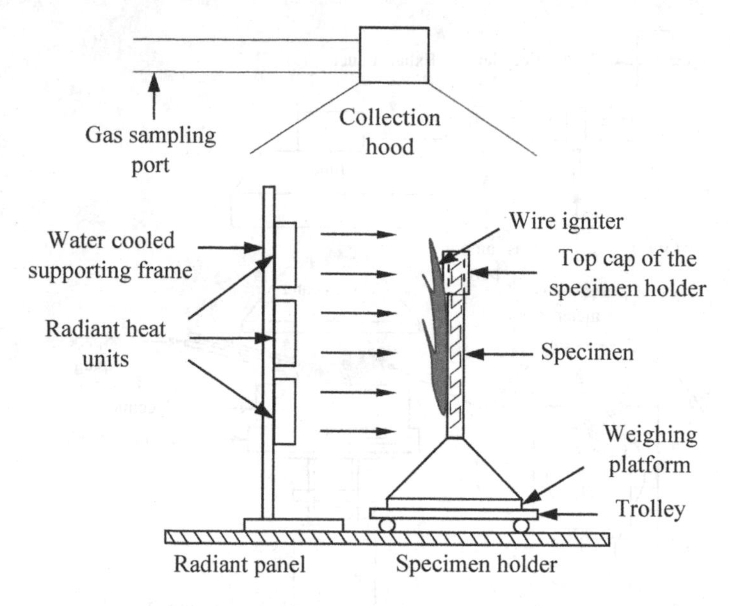

Fig. 1.5 Schematic of an intermediate scale fire propagation apparatus (Specimen size: 1 m × 1 m) [4]

is obtained by using factors such as ignition temperature, critical and external heat flux values, fraction of the critical heat flux that is lost to the ambient, and material thickness.

Heat of gasification (or pyrolysis) is determined using FPA using flaming and non-flaming approaches. Here, the mass loss rate is measured along with the values of heat flux to the surface (convective and radiative), re-radiation loss, and external heat flux, if any. Using these, the heat of gasification is determined. Heat of gasification has also been correlated, as a function of molecular mass, for several hydrocarbons having molecular mass in the range of 30 kg/kmol–250 kg/kmol.

Flame heat flux is measured using standard procedures. Contributions of heat flux to the surface by convection and radiation are determined for various materials with varying surface areas burning in an ambience having varying oxygen concentrations. Here, measurements of mass loss rates at oxygen mass fraction in the range of 0.12–0.6, under air co-flow conditions are carried out.

The convective heat transfer coefficient for the flow field in FPA is determined using combustion of simple fuel such as methanol. As the oxygen mass fraction is increased more than around 0.3, asymptotic behaviors in mass loss rate and flame heat flux are observed. Also, as the material surface area is increased, the radiative contribution of flame heat flux increases and reaches an asymptotic value and convective contribution decreases. Flame spread rate is commonly determined by FPA and a device called Lateral Ignition and Flame spread Test (LIFT) apparatus, similar to that shown in Fig. 1.5. Quantities such as location of pyrolysis front (the length until which pyrolysis has happened in the surface of the solid fuel), flame spread rate,

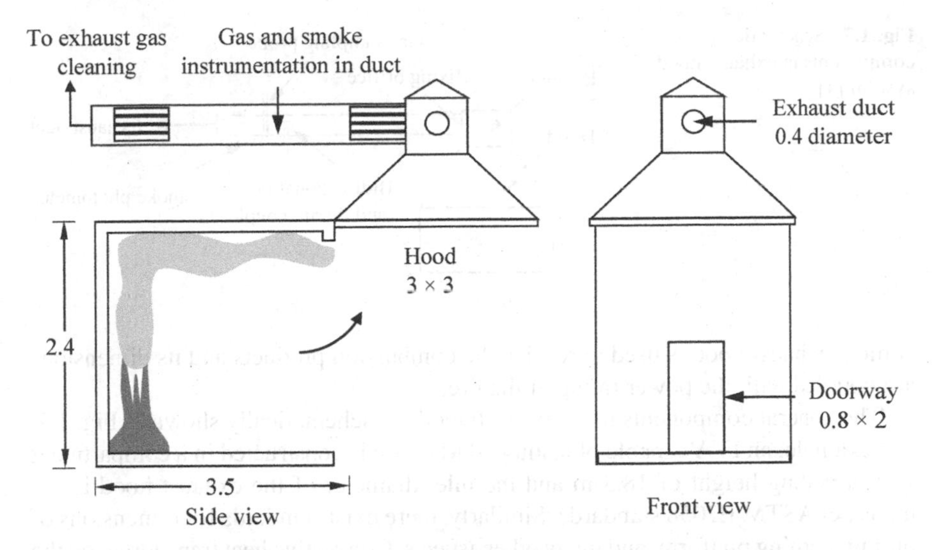

Fig. 1.6 Schematic of standard compartment fire test facility (Dimensions in m) [4]

flame height, and heat release rate as a function of time are measured under various conditions.

Empirical relations for flame height as a function of pyrolysis front, normalized heat release rate, and fire propagation rate are proposed. Fire propagation index (FPI) is correlated in terms of heat release rate and TRP. Flame propagation apparatus is used to determine FPI of various materials by conducting concurrent flame spread tests, considering the aspects of flame radiation, to categorize these materials. Based on the value of FPI, it can be determined whether flame will propagate over the material or not, under a given ambient condition.

Heat release rate, called chemical heat release rate, is measured commonly using techniques such as Carbon Dioxide Generation (CDG) calorimetry and Oxygen Calorimetry (OC). In CDG calorimetry, precise measurements of mass burning rate and rates of generation (kg/m^2-s) of carbon-monoxide (CO) and carbon dioxide (CO_2) are made. For a given fuel, the net heat of complete combustion per unit mass of CO_2 generated and the net heat of complete combustion per unit mass of CO generated can be calculated. By measuring the rates of generation of CO_2 and CO, the heat release rate can be determined. For majority of the fuels, the average value of net heat of complete combustion per unit mass of CO_2 generated is 13.3 ± 1.5 MJ/kg and the average value of the net heat of complete combustion per unit mass of CO generated is 11.1 ± 2 MJ/kg. Similarly, in oxygen calorimetry, precise measurements of mass burning rate and mass consumption rate of oxygen are made. For several fuels, the net heat of complete combustion per unit mass of oxygen consumed is 13.1 ± 0.7 MJ/kg. For several fuels, these values have been tabulated in SFPE Handbook of Fire Protection Engineering [4]. There exists a standard setup for testing a compartment fire. The schematic of the primary components in that setup is shown in Fig. 1.6. A

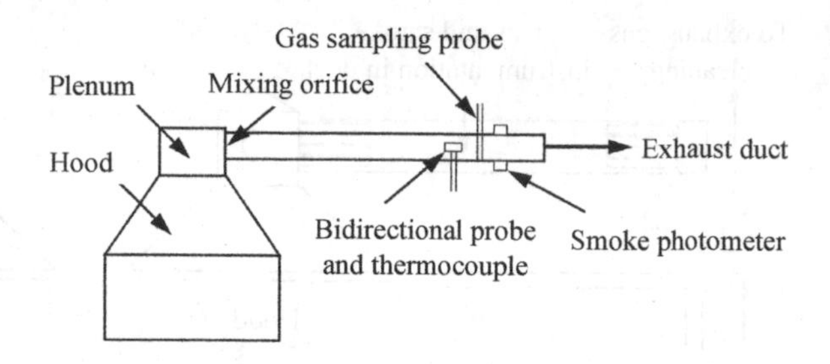

Fig. 1.7 Schematic of components in exhaust hood system [4]

conical exhaust hood is used to receive the combustion products and its dimensions are coupled with the power rating of the fire.

The general components in an exhaust hood are schematically shown in Fig. 1.7. For example, an FPA capable of testing 10 MW fire is constructed in a compartment with a ceiling height of 18.3 m and the inlet diameter of the exhaust hood is 6.1 m, as per ASTM E2058 standards. Similarly, there exist standards for dimensions of piping, burning platform, and many other aspects. Convective heat transported by the product gases is estimated by Gas Temperature Rise (GTR) calorimetry. Here, the temperature difference between the exhaust and ambient gas mixture is measured.

The mass flow rate of the product gases and its specific heat (product mixture) are used to determine the convective heat release. The difference between the net heat release and convective heat release is calculated as the radiative heat release. One more useful measurement that can be done using calorimetry is the smoke point, which is the minimum laminar volumetric flow rate of the fuel at which smoke just escapes from the flame tip. Smoke point values have been reported for several gas, liquid, and solid fuels [4]. Combustion efficiency and CO generation efficiency are correlated to the smoke point. Apart from these, corrosion hazards due to fire can be estimated using this apparatus.

1.5 Non-dimensional Numbers for Fires

Fire can be as small as a room heating fire to as large as a forest fire. Fire hazard in a given structure cannot be easily understood just by constructing a scaled down model of the structure. There are several aspects related to the scaling procedure. A chapter on scaling and dimensional groups related to fire modeling is presented by Quintiere [5] and a discussion on the same by Quintiere et al. [6]. In this section, a brief report on non-dimensional numbers used for fire scaling studies is presented without the involved derivations.

From conservation equations of mass, momentum, energy, and species used to model incompressible reactive flows, several non-dimensional groups can be identified using Buckingham's Pi theorem as shown in Table 1.1, with the nomenclature described in Table 1.2. Let U_∞, p_∞, T_∞, and L be the reference velocity, pressure,

Table 1.1 Non-dimensional groups for fires

Group	Definition/Nomenclature	Equation
$\prod_1$	Mass conservation	$U_\infty t_R/L$
$\prod_2$	Reynolds number	$\rho_\infty U_\infty L/\mu$
$\prod_3$	Froude number	$(U_\infty)^2/(gL)$
$\prod_4$	Euler number	$p_\infty/[\rho_\infty(U_\infty)^2]$
$\prod_5$	Prandtl number	$\mu C/k$
$\prod_6$	Radiation energy/Convective thermal energy	$\sigma T_\infty^3/[\rho_\infty C(gL)^{0.5}]$
$\prod_7$	Physical length/Radiation absorption length	κL
$\prod_8$	Potential energy/Enthalpy	$gL/(CT_\infty)$
$\prod_9$	Fire power/Enthalpy flow rate (Zukoski number)	$\dot{Q}/[\rho_\infty C T_\infty L^2(gL)^{0.5}]$
$\prod_{10}$	Schmidt number	$\mu/(\rho_\infty D)$
$\prod_{11}$	Fuel flow rate/Convective flow rate of mixture	$\dot{m}_{\text{fuel}}/[\rho_\infty L^2(gL)^{0.5}]$
$\prod_{12}$	Transient conduction normal to the wall surface	$[\rho_w C_w \delta_w^2]/k_w t_R$
$\prod_{13}$	Biot number	$h\delta_w/k_w$
$\prod_{14}$	Surface radiation/Conduction	$\varepsilon\sigma T_\infty^3\delta_w/k_w$

Table 1.2 Nomenclature

Symbol	Definition	Units
t_R	Reference time	s
μ	Dynamic viscosity	Pa-s
ρ_∞	Reference density	kg/m^3
C	Specific heat	J/kg-K
k	Thermal conductivity	W/m-K
σ	Stefan-Boltzmann constant	W/m^2-K^4
κ	Absorption coefficient	m^{-1}
$\dot{Q}$	Heat release rate ($\dot{m}''' \times \Delta h_c \times$ Volume)	W
$\dot{m}'''$	Fuel consumption rate per unit volume	kg/m^3-s
Δh_c	Heat of combustion	J/kg
D	Mass diffusivity (Diffusion coefficient)	m^2/s
$\dot{m}_{\text{fuel}}$	Mass flow rate of the fuel gases (pyrolysate)	kg/s
δ_w	Wall thickness (w - wall)	m

temperature, and length, respectively. The non-dimensional group that is obtained from continuity (mass conservation) equation is $\prod_1$. When $\prod_1$ is unity, the reference time (t_R) is written as,

$$t_R = L/U_\infty$$

From the conservation of momentum, the non-dimensional groups of $\prod_2$, $\prod_3$, and $\prod_4$ are defined. When Froude number ($\prod_3$) is unity, the velocity appropriate to buoyancy driven flow and related reference time are obtained. That is,

$$U_\infty = (gL)^{0.5} \quad \text{and} \quad t_R = (L/g)^{0.5}$$

From conservation of energy, five non-dimensional groups are identified namely, $\prod_5$, $\prod_6$, $\prod_7$, $\prod_8$, and $\prod_9$. From conservation of species, two non-dimensional groups are obtained as $\prod_{10}$, and $\prod_{11}$.

Heat transfer from fire to walls (solid surfaces) and conduction heat transfer within the solids are important in modeling several fire scenarios. Considering transient conduction in the direction normal to the wall surface, from the governing equation, non-dimensional group $\prod_{12}$ is obtained. Thermal boundary condition at the wall is written as heat conducted into the wall and is contributed by convection and radiation heat transfer at the wall surface. Further, using the definition of Nusselt number ($\mathrm{Nu} = hL/k$), expressing Nu as $\mathrm{Re}^{0.8} \times \mathrm{Pr}^{0.33}$ (for turbulent flow), the non-dimensional groups $\prod_{13}$, and $\prod_{14}$ are obtained.

When fire tests in scaled down facility are performed, it is expected to preserve several non-dimensional groups between the actual and scaled models simultaneously. This is challenging and therefore, only a few important non-dimensional groups are preserved. Commonly used non-dimensional groups are $\prod_1$, $\prod_2$, $\prod_3$, $\prod_9$, $\prod_{11}$, $\prod_{12}$, and $\prod_{13}$.

General aspects considered are: All variables are made non-dimensional using appropriate reference values. Re is maintained as high as in a turbulent flow. Under natural convective conditions, *Grashof number* ($\sim\mathrm{Re}^{0.5}$) is considered in turbulent regime. Non-dimensional parameters for fuel in scaled model and actual scenario are preserved for required chemical and thermal properties. Fire power and flow rates are also scaled using appropriate non-dimensional groups.

Review Questions

1. How is a fire caused?
2. List different types of combustible materials.
3. What are the classes of fire?
4. Name important thermo-physical properties of materials related to fire analysis.
5. What are the important properties associated with liquid fuels?
6. How are solid fuels classified? What are the important properties associated with these fuels?
7. A rectangular MMA pool, 300 mm in length, 100 mm wide, and 5 mm depth is subjected to a radiant heat flux of 100 kW/m^2 at its surface (300 mm $\times$ 100

mm) and the bottom surface is exposed to ambient (300 K), all other sides are insulated. The boiling point of MMA is 250 °C. Its thermal conductivity, density, and specific heat are 0.147 W/m-K, 940 kg/m^3 and 1.191 kJ/kg-K, respectively. Calculate the time taken to reach the boiling point. Assume heat loss is taking place at an average temperature of 125 °C.

8. Categorize the following quenching agents based on the class of the fire sources it can extinguish: (a) Carbon dioxide, (b) pressurized water, (c) alkaline mixture, and (d) dry chemicals
9. List the typical constituents of combustible solid materials.
10. Compare the rate limiting processes involved in liquid and solid fuels.
11. Define the following: (a) ignition temperature, (b) critical heat flux, (c) re-radiation loss, and (d) thermal response parameter.
12. Hydrogen cylinder(s) leaks into a room of size 5 × 5 × 10 m^3 at 300 K and 1 atm pressure. Calculate the mass of hydrogen required to form a flammable mixture. Assuming that the leak persists for a sufficiently long time, what must be thc mass of hydrogen in the room to just form a non-flammable mixture? The lean and rich flammability limits of H_2 - air mixture (in terms of equivalence ratio) are 0.1 and 7.14, respectively.
13. Hot butane vapors of mass 10 kg are present in a closed room of dimension 4 × 4 × 6 m^3. If a window of the room is opened suddenly, will backdraft occur? The flammability limits of butane (in terms of equivalence ratio) are 0.5 and 2.45, respectively.
14. What are the factors important for piloted ignition?
15. What is autoignition?
16. Define flame heat flux and heat release parameter.
17. List the factors governing flame spread and mass burning rate.
18. Liquid methyl methacrylate (MMA) pool of diameter 25 mm is ignited such that a diffusion flame prevails over it. The thermal power of the flame is 0.1 kW. The enthalpy of vaporization is 333.033 kJ/kg. The heat of combustion is 25527 kJ/kg. Assuming 20% radiation loss to the ambient and 3% conduction loss to the burner walls, calculate the mass burning rate of the fuel. What is the heat flux to the fuel surface? (*Note:* assume steady-state burning for the calculations).
19. List the basic apparatus used for flammability tests.
20. How is flame propagation index correlated?
21. Discuss carbon dioxide generation calorimetry and oxygen calorimetry.
22. Define Froude number and Zukoski number.
23. Propane is supplied at a velocity of 0.01 m/s. The flame length is 30 mm. Evaluate the Froude number of the flame.
24. What are the differences between Nusselt and Biot numbers?
25. Write the non-dimensional group involved with fuel flow rate scaling.
26. Write five commonly preserved non-dimensional groups for fire scaling.
27. The ratio of reference lengths of a model and a prototype is 1 : λ. By preserving $\prod_3$, and $\prod_9$, determine the scaling law for the reference velocity, the reference time, and the heat release rate.

References

1. Rein, G., N. Cleaver, C. Ashton, P. Pironi, and J.L. Torero. 2008. The Severity of Smouldering Peat Fires and Damage to the Forest Soil. *Catena* 74 (3): 304–309.
2. Liang, Y. 2016. *From Shelters to Long Living Communities*, Masters Thesis, University of Massachusetts Amherst.
3. American Society for Testing and Materials: Fire and flammability standards. https://www.astm.org/Standards/fire-and-flammabilitystandards.html.
4. Hurley, M. J., Editor-in-Chief. 2016. *SFPE handbook of fire protection engineering*, Fifth Edition. Springer.
5. Quintiere, J. G. 2006. *Fundamentals of Fire Phenomena*. Wiley.
6. Quintiere, J. G., A. C. Carey, L. Reeves, and L. K. McCarthy. 2017. Scale Modeling in Fire Reconstruction, Document no. 250920, Award no. 2008-DN-BXK178, U. S. Department of Justice.

Chapter 2
Review of Thermodynamics and Transport Processes for Fires

In this chapter, concepts in stoichiometry, chemical thermodynamics, and mass transfer, useful for studying fire, are reviewed. Stoichiometry deals with the calculation of the required amount of oxidizer for completely burning a given amount of fuel. Heat and temperature calculations are carried out using the first law of thermodynamics considering multi-component reactant and product mixtures. Heat of reaction (heat released during combustion) and adiabatic flame temperature (the maximum temperature attained) are important quantities, which quantify fire hazard. Mass transfer basics are required to understand the release and transport of gaseous fuels from condensed fuels such as liquid and solid fuels.

2.1 Stoichiometry

Fuels contributing to fire are mixtures of saturated and unsaturated hydrocarbons, which contain some amount of C and H atoms. In many of these fuels, certain number of N, S, and O atoms are also present. Stoichiometry is the science used to determine the amount of oxidizer (air) required to completely burn one kg of the fuel, to form carbon dioxide and water vapor (called major products of combustion reaction). This is a theoretical value. For this calculation, air is considered to be composed of 21% oxygen and 79% nitrogen by volume, and impurities such as argon and carbon dioxide, usually present in small amounts in atmospheric air, are not considered.

For a general hydrocarbon fuel represented by C_xH_y, containing x carbon atoms and y hydrogen atoms, a general single-step reaction considering its complete combustion in air is written as

$$C_xH_y + a(O_2 + 3.76N_2) \rightarrow xCO_2 + \frac{y}{2}H_2O + 3.76aN_2. \quad (2.1)$$

A. S. Rangwala and V. Raghavan, *Mechanism of Fires*,
https://doi.org/10.1007/978-3-030-75498-3_2

The value of "a" in Eq. (2.1) is obtained by balancing the numbers of carbon, hydrogen, and oxygen atoms on both sides of the reaction, and is given as, $a = x + y/4$. For instance, for methane (CH_4), $x = 1$, $y = 4$, and $a = 1 + 4/4 = 2$. The stoichiometric air-fuel ratio for methane combustion with air is calculated as the mass of air required to completely burn one kg of methane. It is equal to $2 \times 4.76 \times 28.84/16 = 17.16$ kg. And hence, the stoichiometric-air fuel ratio equals 17.16 kg/1 kg = 17.16. Here, 28.84 is molecular weight of air, calculated assuming air to be constituted by 21% oxygen and 79% nitrogen by volume. The molecular weight of air is calculated as $0.21 \times 32 + 0.79 \times 28 = 28.84$ kg/kmol and the molecular weight of methane is 16 kg/kmol. In general, for higher order straight chain hydrocarbons, the stoichiometric air fuel ratio can be roughly taken as 15; that is, for one kg of fuel, approximately 15 kg of air is required theoretically.

It should be kept in mind that in Eq. (2.1) nitrogen is considered to be chemically inert; that it remains as nitrogen in product mixture at the temperature of products. In reality, nitrogen can form its oxides such as NO, NO_2, and N_2O in small amounts. Also, CO_2 and H_2O, at higher temperatures, dissociate to form CO, H_2, OH, and so on. For calculation of stoichiometric (theoretical) air, which is required for heat and temperature calculations, complete combustion as in Eq. (2.1) is assumed.

When the amount of air available for combustion is less than "a" kmol (oxygen starved), species like CO, OH, H, and so on are seen in the product mixture as a result of incomplete combustion. Oxygen starved combustion occurs in several fire scenarios occurring within compartments. In the cases of oxygen starved combustion and when dissociation of the products occurs at high temperature, a knowledge of second law of thermodynamics is required to determine the possible species, which contribute to the product mixture, at a given pressure and temperature. If the air availability is higher than "a" kmol, the excess oxygen is seen in the product mixture. In open fires or fires burning with good ventilation, excess air entrains into the fire and the products contain excess oxygen. A useful parameter, called *equivalence ratio*, is defined as the ratio of stoichiometric air-fuel ratio to the actual air-fuel ratio.

Consider combustion of n-heptane (C_7H_{16}). Theoretically, the number of moles of oxygen required to burn 1 kmol of n-heptane completely is $7 + 16/4 = 11$. If a mixture with more oxygen is considered, say, 13.75 kmol, the reaction is written as

$$C_7H_{16} + 13.75(O_2 + 3.76N_2) \rightarrow 7CO_2 + 8H_2O + 2.75O_2 + 51.7N_2.$$

The excess 2.75 kmol of oxygen is seen in the products. In addition to this, if dissociation of CO_2 occurs due to high temperature, products will contain CO as well. Dissociation of CO_2 occurs well before the dissociation of H_2O. Such a reaction is represented by

$$C_7H_{16} + 13.75(O_2 + 3.76N_2) \rightarrow aCO_2 + bCO + 8H_2O + cO_2 + 51.7N_2.$$

Here, the values of a, b, and c can be determined by invoking a concept called *chemical equilibrium*, which is based on second law of thermodynamics. At a given pressure and temperature, a parameter called *equilibrium constant*, K_p, is evaluated at the given temperature using Gibbs free energy, $g = h - Ts$, applied to a given *elementary reaction*. For example, K_p can be applied to an elementary reversible reaction for CO_2 dissociation, given as $CO_2 \leftrightarrow CO + 0.5\,O_2$. By doing so, the extent of dissociation of CO_2 at the given temperature and pressure can be determined, which will provide the values of a, b, and c. Similarly, molecular level elementary reactions are defined for H_2O dissociation ($H_2O \leftrightarrow H_2 + 0.5\,O_2$), water gas shift reaction ($CO + H_2O \leftrightarrow CO_2 + H_2$), and so on. More details on the evaluation of equilibrium products at a given temperature and pressure are available in basic combustion textbooks [1]. The equilibrium products may also be evaluated using programs such as GASEQ and NASA CEA. The website for GASEQ is http://www.gaseq.co.uk/ and that for CEA is https://cearun.grc.nasa.gov/. It should be noted that these elementary reactions occur at *molecular* level, as opposed to the single-step reaction shown in Eq. (2.1), which is written at *molar* level.

Furthermore, one-step reaction shown in Eq. (2.1) is only a representation of major reactants and products. In reality, a fuel species disintegrates to lower order species and several intermediate species, called *radicals*, are formed during combustion. Even the simplest hydrocarbon, such as methane, disintegrates as follows:

$$CH_4 \rightarrow CH_3 \rightarrow CH_2 \rightarrow CH$$

Starting with CH_4, even C_2 hydrocarbons such as C_2H_2 and C_2H_4 are formed during the oxidation of CH_4. Therefore, several species and reactions are required to thoroughly understand the mechanism by which the products are formed and the rate by which they are formed. However, for heat calculations in most fire safety applications, a single-step reaction is sufficient.

Fire scenarios encounter fuels, which are multi-component in nature. Consider Liquefied Petroleum Gas (LPG), a fossil fuel. It is a multi-component fuel that, on an average basis, typically contains 0.03% CH_4, 0.96% C_2H_6, 13.31% C_3H_8, 10.22% C_3H_6, 30.23% i-C_4H_{10}, 25.32% n-C_4H_{10}, 3.98% C_4H_8, 5.03% i-C_4H_8, 4.99% *trans*-2-C_4H_8, 3.64% *cis*-2-C_4H_8, 1.96% i-C_5H_{12}, and 0.33% n-C_5H_{12}, by volume. It is observed here that isomers of butane such as *i*-butane, and that of butene, such as *i*-butene, *trans*-2-butene, and *cis*-2-butene are present in LPG. The chemical formula for isomers is the same as normal species, but, its chemical structure varies. Since volume percentages of the constituents are given, these would convert as mole percentages. Therefore, for 100 kmol of fuel, the single-step reaction can be written as

$$\begin{aligned}
&0.03CH_4 + 0.96C_2H_6 + 13.31C_3H_8 + 10.22C_3H_6 + (30.23 + 25.32)C_4H_{10} \\
&+(3.98 + 5.03 + 4.99 + 3.64)C_4H_8 + (1.96 + 0.33)C_5H_{12} \\
&+ m(O_2+3.76N_2) \rightarrow nCO_2 + pH_2O + 3.76mN_2.
\end{aligned}$$

Or for one kmol of fuel,

$$0.0003CH_4 + 0.0096C_2H_6 + 0.1331C_3H_8 + 0.1022C_3H_6 + 0.5555C_4H_{10}$$
$$0.1764C_4H_8 + 0.0229C_5H_{12} + a(O_2 + 3.76N_2) \rightarrow bCO_2 + cH_2O + 3.76aN_2.$$

Here, a, b, c, and d are determined by carbon, hydrogen, and oxygen balances. Alternatively, LPG can be written as a single fuel with a structure of C_xH_y, where x and y are determined by the percentages of its constituents. By this approach, LPG can be considered as $C_{3.7675}H_{8.9778}$. For this fuel, the stoichiometric air is determined by writing the single-step reaction as,

$$C_{3.77}H_{8.98} + a(O_2 + 3.76N_2) \rightarrow bCO_2 + cH_2O + 3.76aN_2$$

Here, by balancing the atoms, $b = 3.77$, $c = 4.49$, and $a = 6.015$. The molecular weight of LPG is $12 \times 3.77 + 8.98 \times 1 = 54.2$ kg/kmol. For burning one kg of LPG completely, theoretically, 15.23 kg of air is required.

Consider a solid fuel such as wood. Results from proximate analysis of wood give the weight percentages of its volatile, fixed carbon, moisture, and ash contents. Its ultimate analysis provides the elemental composition such as percentages of C, H, O, and N, on mass basis. When heated to a certain temperature, gaseous fuels, called volatiles, trapped inside the wood are released. Consider a typical wood that has 80% volatiles and 20% fixed carbon from proximate analysis and 50% C, 8% H, 41.5% O, and 0.5% N, obtained from its ultimate analysis, both on dry (moisture free) and ash free basis. It should be noted that carbon contained in volatiles contributes to the gas-phase reactions, which are generally more rapid. Carbon in solid form burns slowly due to the reaction occurring at its surface.

When the solid fuel is heated in an inert (containing no oxygen) atmosphere, volatiles are released, and fixed carbon and ash remain as solid residuals. For calculating the stoichiometric air required for gas-phase reactions involving volatiles alone, the fuel is represented in a consolidated form, $C_xH_yO_zN_p$. The moles of C, H, O, and N, which are x, y, z, and p, are evaluated considering results from proximate and ultimate analyses. For this calculation, molecular weight of gaseous mixture constituting the volatile matter is also required. Typical volatile species trapped in a solid fuel are CH_4, C_2H_2, C_2H_4, CO, H_2, O_2, and N_2, in some proportions, and the typical molecular weight of volatile mixture is around 30 kg/kmol. The value of x in the consolidated fuel is calculated as

$$\frac{\text{(Weight of carbon in volatile/Molecular weight of carbon)}}{\text{(Weight of volatile in wood/Molecular weight of volatile)}}$$

For this case, out of 50% C, 20% is in solid form, therefore, x in $C_xH_yO_zN_p$ is evaluated as

$$x = \frac{(50-20)/12}{80/30} = 0.9375.$$

Noting that hydrogen is present only in volatiles, the value of y is obtained as

$$y = \frac{8/1}{80/30} = 3.$$

Similarly, the values of z and p are obtained as

$$z = \frac{41.5/16}{80/30} = 0.9726, \qquad p = \frac{0.5/14}{80/30} = 0.0134.$$

The volatile is now written as $C_{0.94}H_3O_{0.97}N_{0.013}$. It is customary to rewrite this as $CH_{3.2}O_{1.03}N_{0.014}$, by keeping the number of atoms of C equal to unity. The molecular weight of this fuel is 31.88 kg/kmol. Further, N can be neglected and the number of atoms of H and O can be rounded off such that the fuel is represented as CH_3O. Its molecular weight is 31 kg/kmol, showing only a 2.8% reduction, which is good enough for fire analysis. The stoichiometric air required for its complete combustion can be determined. The single-step reaction is written as

$$CH_3O + 1.25(O_2 + 3.76N_2) \rightarrow CO_2 + 1.5H_2O + 4.7N_2. \tag{2.2}$$

For this case, the stoichiometric air-fuel ratio for burning the volatiles is 5.535 kg/kg-volatile. It should be noted that the oxygen inherent in the fuel contributes to a notable portion for combustion. When total air required for burning unit mass of the wood (including both volatiles and fixed carbon) is calculated, the unknowns in the consolidated fuel ($C_xH_yO_zN_p$) are evaluated from the ultimate analysis only. For the case discussed above, $x = 0.5/12$, $y = 0.08/1$, $z = 0.415/16$, and $p = 0.005/14$. The consolidated fuel having one carbon atom and neglecting N is written as $CH_{1.92}O_{0.63}$. The single-step reaction is written as

$$CH_{1.92}O_{0.63} + 1.165(O_2 + 3.76N_2) \rightarrow CO_2 + 0.96H_2O + 4.38N_2. \tag{2.3}$$

The stoichiometric air-fuel ratio for burning the volatiles is 6.66 kg/kg-wood. That is, typically around 6.5 kg air is required to burn 1 kg of dry wood. It is possible to apply the concept of stoichiometry to estimate the type of fuel and air-fuel ratio with the information of the product composition. For example, consider combustion of a saturated hydrocarbon fuel, C_xH_y, and air. The dry product mole fractions are measured as 83.61% N_2, 4.91% O_2, 10.56% CO_2, and 0.92% CO. The composition of the fuel and equivalence ratio of the reactant mixture can be determined as follows.

Since the fuel is a saturated hydrocarbon, $y = 2x + 2$. For producing 100 kmol of the dry products, which excludes H_2O vapor, let "a" kmol of fuel burn. For this, the single-step reaction is written as

$$\begin{aligned} &aC_xH_{2x+2} + b(O_2 + 3.76N_2) \\ &\quad \rightarrow 10.56CO_2 + 0.92CO + cH_2O + 4.91O_2 + 83.61N_2. \end{aligned}$$

By atom balance, $b = 83.61/3.76 = 22.2367$, $c = 44.4734 - 21.12 - 0.92 - 9.82 = 12.61$. By C balance: $a(x) = 10.56 + 0.92 = 11.48$. Also, by H balance, $a(2x +$

2) = $2c = 25.23$. Solving these, the value of $a = 1.13$ and the value of $x = 10.13$. Since the hydrocarbon is saturated, nearest value of x can be taken as 10. Thus, the fuel is $C_{10}H_{22}$, n-decane. Since x is taken as 10, the value of "a" is recalculated as $11.48/10 = 1.148$ in order to keep 100 moles of dry products. Writing the chemical reaction for 1 kmol of fuel,

$$C_{10}H_{22} + 19.37(O_2 + 3.76N_2) \rightarrow 9.1986CO_2 + 0.8014CO + 10.987H_2O + 4.277O_2 + 72.831N_2.$$

For stoichiometric combustion of n-decane and air, the reaction is written as

$$C_{10}H_{22} + 15.5(O_2 + 3.76N_2) \rightarrow 10CO_2 + 11H_2O + 58.28N_2.$$

The stoichiometric air-fuel ratio is $15.5 \times 4.76 \times 28.84/142 = 14.985$. The actual air-fuel ratio is estimated as $19.37 \times 4.76 \times 28.84/142 = 18.726$. The equivalence ratio, ϕ is $15.5/19.37 = 0.8$. This can be represented as $100 \times 19.37/15.5 = 125\%$ stoichiometric air (or 25% excess air). It is noted that percent stoichiometric air is also calculated as $100/\phi$. For air rich (or fuel lean) reactant mixture, as in this case, due to dissociation of CO_2, the product mixture also contains CO in it.

2.2 Heat Calculations

Heat release from a fire is an important quantity to be estimated in fire safety engineering. *Fire loading* is a parameter that indicates the potential hazard of a fire. This is calculated as the heat release flux, that is, the heat output from a fire per unit area (W/m^2). In general, if the calorific or heating value of the material (in J/kg) that is burning is known and the rate at which it is burning (in kg/s) is also known, then heat release rate is calculated as the product of these two. Calorific value is the heat released when unit mass of the material is completely burnt. Calorific values of materials used in several applications are determined by using different types of calorimeters, including the *bomb calorimeter*.

Theoretically, heating value is calculated using the first law of thermodynamics. For this analysis, a control (constant) volume is considered with the fire source (material that is burning) at one of its boundaries. Fuel gases from this source flow into the control volume. Air from the ambient also flows into the control volume and products of combustion flow out of the control volume. Assuming steady state and steady flow for simplicity, the first law of thermodynamics is written for the control volume as

$$\dot{Q} - \dot{W}_x = \sum_1^N \dot{n}_i \overline{h_i}(T_P) - \sum_1^M \dot{n}_i \overline{h_i}(T_R). \tag{2.4}$$

Here, N and M are the number of product species and reactant species, respectively, $\dot{n}$ is the flow rate in kmol/s and T is the temperature in K. Subscripts P and R, represent product and reactant, respectively. The term $\dot{Q}$ is the rate of heat transfer from or to the control volume, based on if it is negative or positive. The term $\overline{h}_i$ is called *standard or absolute enthalpy* of ith species expressed in J/kmol. The term $\dot{W}_x$ is the rate of work interaction of control (constant) volume, and is neglected in this case. Therefore,

$$\dot{Q} = \sum_{1}^{N} \dot{n}_i \overline{h_i}(T_P) - \sum_{1}^{M} \dot{n}_i \overline{h_i}(T_R). \tag{2.5}$$

For example, consider single-step reaction for methane,

$$CH_4 + 2(O_2 + 3.76N_2) \rightarrow CO_2 + 2H_2O + 7.52N_2.$$

It is noted that methane enters the control volume at the rate of 1 kmol/s, oxygen enters the control volume at the rate of 2 kmol/s and CO_2 and H_2O, leave the control volume at the rate of 1 kmol/s and 2 kmol/s, respectively. For this reaction, the rate of heat transfer is calculated as

$$\dot{Q} = \overline{h}_{CO_2}(T_P) + 2\overline{h}_{H_2O}(T_P) + 7.52\overline{h}_{N_2}(T_P) - \overline{h}_{CH_4}(T_R) \\ - 2\overline{h}_{O_2}(T_R) - 7.52\overline{h}_{N_2}(T_R). \tag{2.6}$$

Standard enthalpies of the species are calculated primarily as a function of temperature. As the temperature increases, the value of enthalpy also increases. Only under high-pressure conditions, the pressure dependency for enthalpy would arise. In a combustion reaction, which is an exothermic oxidation reaction, heat is generated because the enthalpy of the product mixture is less than the enthalpy of the reactant mixture. That is, $\dot{Q}$ is negative for an exothermic reaction. In other words, heat is transferred from the control volume to the surroundings. It is further noted that the rate of heat transfer to the surroundings is the maximum, when the temperature of the product is as low as that of the ambient. *Standard heat of reaction* is defined as the heat generated when reactants are supplied at 1 atm, 298 K (reference state) and the products formed are also cooled to 298 K. That is, $\dot{Q}$ is called the standard heat of reaction when the value of both T_P and T_R in Eq. (2.5) is 298 K. The standard enthalpy of species i is expressed in molar basis as

$$\overline{h}_i(T) = \overline{h}^0_{f,i}(T_{\text{ref}}) + \Delta\overline{h}_i(T). \tag{2.7}$$

In Eq. (2.7), the first term on the right-hand side, $\overline{h}^0_{f,i}$, is called the *standard enthalpy of formation* for the given species in kJ/kmol. The overhead bar indicates that the enthalpy is specific to number of moles. The superscript "0" indicates that the value is obtained at 1 atmosphere pressure and at the standard state temperature

(298 K). It is defined as the increase in enthalpy, when one mole of a compound is formed at constant pressure from its natural elements, which are in the standard state (1 *atmosphere*, 298 K) and the compound, after its formation, is also brought to the standard state. For naturally occurring elements such as O_2, H_2, N_2, C (solid), etc., the standard enthalpy of formation is assigned a value of zero. For other species, its formation is considered by the reactions of the above-mentioned basic elements. For example, CO is considered to be formed by the reaction at standard state between C (graphite) and O_2, written as C (s) $+ 0.5O_2 \rightarrow$ CO. Similarly, hydrocarbon fuel species are formed by the reaction between C (graphite) and H_2.

The second term in the right-hand side of Eq. (2.7), $\Delta\overline{h}_i(T)$, is the enthalpy associated with the temperature change for that species. It is called *sensible enthalpy*. This is the increase in the enthalpy due to an increase in the temperature from the standard value (298 K) to a higher temperature, T (K). For several species, the enthalpies of formation, sensible enthalpies, and specific heat values as a function of temperature are available in NIST website (http://webbook.nist.gov/chemistry/). For ideal gases, which are also thermally perfect, enthalpy and internal energies are functions of temperature alone. The sensible enthalpy is calculated by integrating the specific heat at constant pressure, which is generally expressed as a polynomial in temperature.

In order to calculate the value of standard heat of reaction given in Eq. (2.6), the required data is obtained from NIST website. It should be noted that the reactants are entering at $T_{\text{ref}} = 298$ K and the products formed are cooled to 298 K and the term $\Delta\overline{h}_i(T)$ in Eq. (2.7) is zero. Moreover, the enthalpies of formation for O_2 and N_2 are zero. The calculation is illustrated below. The superscript 0 has been dropped for simplicity.

$$\begin{aligned}
\dot{Q} &= \overline{h}_{CO_2}(T_P) + 2\overline{h}_{H_2O}(T_P) + 7.52\overline{h}_{N_2}(T_P) - \overline{h}_{CH_4}(T_R) \\
&\quad - 2\overline{h}_{O_2}(T_R) - 7.52\overline{h}_{N_2}(T_R) \\
\dot{Q} &= \overline{h}_{f,CO_2}(T_{\text{ref}}) + 2\overline{h}_{f,H_2O}(T_{\text{ref}}) + 7.52\overline{h}_{f,N_2}(T_{\text{ref}}) \\
&\quad - \overline{h}_{f,CH_4}(T_{\text{ref}}) - 2\overline{h}_{f,O_2}(T_{\text{ref}}) - 7.52\overline{h}_{f,N_2}(T_{\text{ref}}). \\
\dot{Q} &= \overline{h}_{f,CO_2}(T_{\text{ref}}) + 2\overline{h}_{f,H_2O}(T_{\text{ref}}) - \overline{h}_{f,CH_4}(T_{\text{ref}}).
\end{aligned}$$

Substituting the enthalpies of formation of CO_2, H_2O, and CH_4 from NIST data base, the standard heat of reaction of methane is obtained as,

$$\dot{Q} = -393.52 + 2(-241.83) - (-74.5) = -802.68 \text{ kJ/mol - CH}_4$$

It is clear that the value of standard heat of reaction is negative, since heat is released from the control volume to the surroundings (exothermic reaction). Further, this value is obtained per mole of methane, since the reaction between methane and air is written considering one mole (or kmol) of methane. This is the heat released in kW per unit molar flow rate of methane, expressed as kW/(mol-CH_4/s). This may be converted to per kg of methane by dividing the value by molecular mass of methane. That is, standard heat of reaction per kg of methane is $-802.68 \times 1000/16 = -50167.5$ kJ/kg - CH_4 or -50.2 MJ/kg - CH_4.

Heat of combustion or the heating value or the calorific value is just the negative of the standard heat of reaction. Thus, the heating value of methane is 50.2 MJ/kg - CH_4. In the above calculation, water in the product is taken to be in vapor state (the enthalpy of formation of water vapor has been used). Therefore, the heating value, thus obtained, is called the *lower heating value*. On the other hand, if the water in the products is considered to be in liquid state, with enthalpy of formation value equal to -283.83 kJ/mol, then *higher heating value* is obtained. In reality, some water in the products can condense on walls and surfaces and some remain in vapor state. Therefore, the heat of combustion will be between the higher and lower heating values.

The heat of reaction can be calculated per mole of oxygen consumed, or per mole of CO_2 produced during the combustion. For instance, the heat of reaction is calculated per mol of oxygen, as, $-802.68/2 = -401.34$ kJ/mol-O_2, since for one mole of methane burnt, 2 moles of oxygen are burnt. The heat of reaction on oxygen mass basis is: $-401.34 \times 10^3/32 = -12542$ kJ/kg-O_2. If the same calculations are carried out for methanol, the values of heat of combustion are obtained as 19.9 MJ/kg-methanol and 13.3 MJ/kg-O_2. In fact, it can be shown that the heat of combustion per kg-oxygen, for several fire sources (fuels) lies within 13.1 ± 0.7 MJ/kg-O_2. Similarly, the heat of combustion per kg-CO_2 produced is: $802.68 \times 10^3/44 = 18243$ kJ/kg-CO_2. In fact, this is an extreme value. For pentane to dodecane, this value lies in the range of 14.7 MJ/kg-CO_2 to 14.2 MJ/kg-CO_2. The average value of heat of combustion per kg-CO_2 for several fuels is 13.3 ± 1.5 MJ/kg-CO_2, as reported in Chap. 1. The stoichiometric combustion of n-heptane (C_7H_{16}), a commonly used liquid fuel, is given as,

$$C_7H_{16} + 11(O_2 + 3.76N_2) \rightarrow 7CO_2 + 8H_2O + 41.36N_2.$$

The standard enthalpy of formation of gaseous n-heptane is -189.3 kJ/mol and that of liquid n-heptane is -225.9 kJ/mol. Based on the state of the fuel, liquid or vapor, appropriate value of enthalpy of formation is to be used. The standard heat of reaction is calculated when n-heptane exits in vapor phase as,

$$\dot{Q} = 7(-393.52) + 8(-241.83) - (-189.3) = -4500 \text{ kJ/mol - } C_7H_{16}.$$

On mass basis, the heat of reaction is (-4500 kJ/mol$\times$1000)/(100 kg/kmol) = -45000 kJ/kg-fuel. The lower calorific value of n-heptane is 45 MJ/kg, when n-heptane exits in vapor phase. The lower heating value of n-heptane, if it is in liquid state initially, is 44.6 MJ/kg. Standard heat of combustion is obtained when reactant enter and products leave at standard conditions of 1 atm and 298 K. If the products exit the control volume at a higher temperature, the heat of combustion will be lower. For example, consider n-heptane burning with 125% stoichiometric air, written as

$$C_7H_{16} + 13.75(O_2 + 3.76N_2) \rightarrow 7CO_2 + 8H_2O + 2.75O_2 + 51.7N_2.$$

If the reactants enter at 298 K and the products leave at 1200 K, the heat of reaction is calculated as follows:

$$\dot{Q} = 7(-393.52+\Delta\overline{h}_{CO_2}(1200)) + 8(-241.83 + \Delta\overline{h}_{H_2O}(1200)) + 2.75(\Delta\overline{h}_{O_2}(1200)) + 51.7(\Delta\overline{h}_{N_2}(1200)) - (-189.3).$$

It should be noted that since products are at a temperature higher than 298 K, sensible enthalpies [term $\Delta\overline{h}_i(T)$ in Eq. (2.7)] of the species are added appropriately. The enthalpies of formation of O_2 and N_2 are zero and have not been included. The sensible enthalpies are obtained from tabulated data or from specific heat data that is usually available as a piece-wise polynomial (NIST database) as shown below:

$$\Delta\overline{h}_i(1200) = \int_{298}^{1200} C_{p,i}(T)dT = 1000\int_{298}^{1200} (A + BT^* + CT^{*2} + DT^{*3} + ET^{*-2})dT^*.$$

Here, $T^* = (T/1000)$. For instance, considering $i = CO_2$, $A = 24.99735$, $B = 55.18696$, $C = -33.69137, D = 7.948387, E = -0.136638$ (NIST database).

$$\Delta\overline{h}_{CO_2}(1200) = 1000 \times \left(AT^* + \frac{BT^{*2}}{2} + \frac{CT^{*3}}{3} + \frac{DT^{*4}}{4} - \frac{E}{T^*}\right)\Bigg|_{298}^{1200}$$

$$= 10^3\Bigg(24.99735(1.2 - 0.298) + \frac{55.18696(1.2^2 - 0.298^2)}{2} - \frac{33.69137(1.2^3 - 0.298^3)}{3} + \frac{7.948387(1.2^4 - 0.298^4)}{4} + \frac{0.136638}{1.2} - \frac{0.136638}{0.298}\Bigg) = 44.47 \text{ kJ/mol}.$$

Similarly, substituting all the values,

$$\dot{Q} = 7[-393.52 + 44.47] + 8[-241.83 + 34.51] + 2.75(29.7) + 51.7(28.11) - (-189.3) = -2377.65 \text{ kJ/mol - } C_7H_{16}.$$

It is clear that the excess oxygen and nitrogen absorb heat from the fire and increase their sensible heats.

As seen earlier, consider LPG, a multi-component fuel, that typically contains 0.03% CH_4, 0.96% C_2H_6, 13.31% C_3H_8, 10.22% C_3H_6, 30.23% *i*-C_4H_{10}, 25.32% *n*-C_4H_{10}, 3.98% C_4H_8, 5.03% *i*-C_4H_8, 4.99% *trans*-2-C_4H_8, 3.64% *cis*-2-C_4H_8, 1.96% *i*-C_5H_{12}, and 0.33% *n*-C_5H_{12}, by volume. The volume fractions translate directly to mole fractions. That is, the mole fraction of propane, $\chi_{propane}$, in LPG is 0.1331. The molecular weight of LPG is obtained by a chain mixing rule called Kay's rule, given as, $MW_{LPG} = \Sigma \chi_i \times MW_i$. Here, χ_i is the mole fraction of the ith species MW_i is its molecular mass. The molecular mass of LPG is obtained as 54.2 kg/kmol. The standard enthalpy of formation of LPG is obtained using similar chain rule using the enthalpies of formation of the individual species constituting LPG. It is written as

$$\overline{h}_{f,LPG} = \Sigma(\chi_i \times \overline{h}^0_{f,i}) \text{ kJ/mol}. \quad (2.8)$$

It should be noted that if mass fractions of the constituents of LPG have been given, then, the values of enthalpies of formation of individual species used in the calculations should be specific to mass (kJ/kg). These are obtained by dividing the molar enthalpy of formation (kJ/kmol) of a component by its molecular mass. The mixture enthalpy of formation (for LPG) may be converted back to the molar form by using the molecular mass of the mixture, MW_{LPG}. The global single-step reaction of LPG and air, as shown earlier, is written as

$$C_{3.77}H_{8.98} + 6.015(O_2 + 3.76N_2) \rightarrow 3.77CO_2 + 4.49H_2O + 22.6N_2.$$

The standard heat of reaction of LPG and air is calculated as

$$\dot{Q} = 3.77\overline{h}_{f,CO_2}(298) + 4.49\overline{h}_{H_2O}(298) - \overline{h}_{f,LPG}(298)$$
$$\dot{Q} = 3.77(-393.52) + 4.49(-241.83) - (-90.3) = -2479.1 \text{ kJ/mol-LPG}.$$

The lower heating value of LPG is evaluated as 2479.1 × 1000/54.2 = 45740 kJ/kg-LPG.

For heterogeneous fuels such as wood, it is non-trivial to determine the enthalpy of formation. As discussed earlier, wood or similar fuels are characterized by ultimate and proximate analyses. These reveal the percentages of volatiles and fixed carbon (flammable components) and elemental composition of C, H, O, N, and S. As seen earlier, a consolidated fuel in terms of $C_xH_yO_zN_p$ can be formulated and the values of x, y, z, and p can be evaluated using the data from proximate and ultimate analyses. However, for calculating heat of reaction, the enthalpy of formation of the consolidated fuel is necessary. More data is required to arrive at the enthalpy of formation. For example, the composition of volatile in terms of species, such as CH_4, C_2H_2, C_2H_4, CO, H_2, O_2, and N_2, can be determined by heating the solid fuel in an inert environment and determining the proportions of these gases using gas chromatography. From the composition of volatiles, its enthalpy of formation can be determined by using the same procedure as illustrated for LPG above in Eq. (2.8). The single-step reaction for the consolidated fuel (CH_3O) considering volatiles alone has been illustrated in Eq. (2.2). It is written as

$$CH_3O + 1.25(O_2 + 3.76N_2) \rightarrow CO_2 + 1.5H_2O + 4.7N_2.$$

The heat of reaction for the above reaction is calculated as

$$\dot{Q} = \overline{h}_{f,CO_2}(298) + 1.5\overline{h}_{f,H_2O}(298) - \overline{h}_{f,volatile}(298).$$

Since enthalpy of formation of the fuel representing the volatile has been determined by measuring the species constituting the volatiles by gas chromatography, the heat of reaction and the heating value of the volatiles in the solid fuel can be determined. Remaining is the combustion of fixed carbon, $C(s)$, which is represented as

$$C(s) + O_2 \rightarrow CO_2 \text{ (for its complete combustion).}$$

It may be noted that the standard heat of reaction of the above reaction is the enthalpy of formation of CO_2. The combustion of charring solid fuel such as wood can be analyzed by splitting its homogeneous (gas phase) and heterogeneous (solid phase) parts. If the heating value of wood is determined fairly accurately by calorimetry, then it can be used to calculate the enthalpy of formation of the representative consolidated fuel. For this, consider Eq. (2.3):

$$CH_{1.92}O_{0.63} + 1.165(O_2 + 3.76N_2) \rightarrow CO_2 + 0.96H_2O + 4.38N_2.$$

The heat of reaction, which is negative of the heat of combustion or the calorific value, is evaluated as

$$\dot{Q} = \overline{h}_{f,CO_2}(298) + 0.96\overline{h}_{f,H_2O}(298) - \overline{h}_{f,wood}(298).$$

Since the heating value has been measured using calorimetry, the left-hand side of the above equation is known. The unknown in the right-hand side, which is the enthalpy of formation of the fuel representing wood, can be calculated. This is a useful data that can be used in theoretical and numerical models. In this manner, experimental measurements (gas chromatography and calorimetry) complement the heat calculation for heterogeneous and complex fuels.

2.3 Flame Temperature

Temperature of the flame/fire is an important parameter. The maximum temperature is attained by the flame, when there is no heat transfer to the surroundings. Setting the left-hand side term in Eq. (2.5) to zero, the equation is written as

$$0 = \sum_{1}^{N} \dot{n}_i \overline{h}_i(T_P) - \sum_{1}^{M} \dot{n}_i \overline{h}_i(T_R). \tag{2.9}$$

This means that the enthalpy of the reactant mixture, calculated at T_R, is equal to the enthalpy of the product mixture, calculated at T_P, when there is no heat transfer to the surroundings. The temperature T_P is called *adiabatic flame temperature*, T_{ad}, the maximum attainable temperature for the given reactant mixture at T_R. The calculation of enthalpy of reactant mixture is straight forward, as seen in the heat calculations. However, in order to determine the maximum product temperature, the exact proportions of the species constituting the products should be known. This is non-trivial. If complete combustion is assumed, then for a stoichiometric reactant mixture, CO_2, H_2O, and N_2, in some proportions, contribute to the product mixture.

The adiabatic flame temperature calculated by balancing the enthalpies of the reactant and product mixtures for this case is generally found to be much higher than the measured value. This is because, at high temperatures, as previously discussed, dissociation of the major product species takes place. Since dissociation reaction is an endothermic reaction, some amount of heat from the flame is taken by that reaction. Therefore, even without any heat loss to the surroundings, heat from the flame is fed to the dissociation reactions. As a result of this, only when the exact proportions of the species in the product mixture are taken into account, realistic value of adiabatic flame temperature can be obtained. In order to evaluate the exact product species, the concept of chemical equilibrium, as discussed earlier in this chapter is used. The flame temperature calculated using equilibrium approach is found to be closer to the measured values.

Consider char (solid carbon) initially at 298 K burns in air in a furnace. The air flow is measured to be 150% theoretical air. Since this is a case with excess air, the resultant temperature is expected to be low and the chance of dissociation is also small. Therefore, neglecting the dissociation of CO_2 formed, the adiabatic flame temperature is calculated as illustrated below. The combustion reaction is written in the form:

$$C(s) + 1.5(O_2 + 3.76N_2) \rightarrow CO_2 + 0.5O_2 + 5.64N_2. \tag{2.10}$$

The enthalpy of the reactant mixture is zero, as naturally occurring reactants are initially at 298 K. The enthalpy of the product mixture is calculated using the expression,

$$H_P = [\overline{h}_{f,CO_2}(298) + \Delta\overline{h}_{CO_2}(T_P)] + 0.5\Delta\overline{h}_{O_2}(T_P) + 5.64\Delta\overline{h}_{N_2}(T_P). \tag{2.11}$$

Here, $\overline{h}_{f,CO_2}(298)$ is the heat of formation of CO_2 and $\Delta\overline{h}_i$ ($i = CO_2, O_2, N_2$) represents the sensible heat. Since the enthalpy of reactant mixture is zero, the value of T_P at which the value of H_P is zero has to be evaluated. For this, a value of T_P is assumed. In NIST database, the values of standard enthalpies of several species are listed as a polynomial of temperature. Using the polynomial, plugging the assumed value of T_P, H_P is evaluated.

If the value of H_P is close to zero (that is, the difference between H_P and zero is of the order of a small number, say, 10^{-5}), then that value of T_P is the adiabatic flame temperature of the problem under discussion. For this case, when T_P is assumed as 1800 K, the value of H_P is -12.011 kJ. When T_P is assumed as 1900 K, the value of

H_P is 15.9 kJ. Thus, by linear interpolation, the correct value of T_P is estimated as 1843 K. The adiabatic flame temperature is less than 2000 K, since 50% excess air is supplied in this case. Normally at a temperature of around 1900 K, it is expected that about 1% of CO_2 formed will dissociate.

As one more example, consider an adiabatic flow reactor in which vaporized benzene (C_6H_6) and stoichiometric air flow steadily at 1 atm, 298 K, and the products of combustion leaves the reactor.

$$C_6H_6 + 7.5(O_2 + 3.76N_2) \rightarrow 6CO_2 + 3H_2O + 28.2N_2. \quad (2.12)$$

If dissociation of CO_2 and H_2O are neglected, the adiabatic flame temperature is obtained, by a similar calculation illustrated above, as 2529 K. If dissociation of the products is considered, using chemical equilibrium approach, the adiabatic flame temperature is evaluated as 2341 K. This calculation is done using NASA CEA. Mole fractions of CO, CO_2, H_2, and H_2O, in the product mixture, at this temperature are evaluated as 0.0215, 0.1374, 0.002, and 0.075, respectively. A notable decrease in temperature is obtained when dissociation is included.

When the previous example of char combustion in 50% excess air, given in Eq. (2.10), is solved using the NASA CEA code, the adiabatic flame temperature is obtained as 1836 K, which is only 7 K less than the value obtained using the assumption of no dissociation of CO_2. In fact, for this case, the mole fractions of CO_2 and CO are obtained as 0.14 and 0.00016, respectively.

A thermodynamic approach can be used to evaluate the flame temperature using the specific heat of the product mixture. However, a simplification is done in this approach. The temperature at which the specific heat of the mixture is calculated is an average temperature. There are several thumb rules available in literature [2], illustrating the value of the average temperature used to obtain the properties such as specific heat. This approach is illustrated subsequently. Consider Eq. (2.12). The enthalpies of reactant and product mixtures, expressing the sensible enthalpy in terms of specific heat at a constant pressure of the mixture, are written as

$$H_R = \overline{h}_{f,Benzene}$$

$$H_P = 6(-393.52) + 3(-241.83) + n_P \overline{c}_{p,mix}(T_P). \quad (2.13)$$

Here, n_P is the total number of product moles, equal to 37.2 as per Eq. (2.12). The molar specific heat at constant pressure of the product mixture is evaluated using the mole fractions of individual species and Kay's rule as given below.

$$\overline{c}_{p,mix}(T_P) = \Sigma(\chi_i \overline{c}_{p,i}) \text{ kJ/mol-K}. \quad (2.14)$$

The mole fractions are evaluated using the individual number of moles of each product species in Eq. (2.12). For CO_2, it is 0.161 and for H_2O, it is 0.081 and for N_2, it is 0.758. The values of molar specific heats of individual species, appearing in the summation term in the right-hand side of Eq. (2.13) are evaluated at an average temperature. It can be recognized that for adiabatic flame temperature, $H_R = H_P$. Therefore, the problem of estimating T_P simplifies to solving the following equation.

$$T_P = (H_R - [6(-393.52) + 3(-241.83)])/(n_P \times \overline{c}_{p,mix}) \tag{2.15}$$

$$T_P = (82.93 - [6(-393.52) + 3(-241.83)])/(n_P \times \overline{c}_{p,mix})$$
$$T_P = (3169)/(n_P \times \overline{c}_{p,mix}).$$

Let 298 K be considered for evaluating $c_{p,mix}$. The values of c_p in kJ/kg-K for CO_2, H_2O, and N_2, evaluated at 298 K are 0.844, 1.93, and 1.04, respectively. The mixture specific heat is calculated as

$$\overline{c}_{p,mix} = \frac{(0.161 \times 0.844 \times 44 + 0.081 \times 1.93 \times 18 + 0.758 \times 1.04 \times 28)}{1000}$$
$$= 0.0309 \text{ kJ/mol-K}.$$

With this value of mixture specific heat, T_P is evaluated as

$$T_P = (3169)/(37.2 \times 0.0309) = 2757 \text{ K}.$$

This is about 225 K higher than the value calculated without considering dissociation (2529 K). Let the specific heats at constant pressure be calculated at 1200 K, which is typically an average temperature in a reactive flow field. The values of c_p in kJ/kg-K, at 1200 K, for CO_2, H_2O, and N_2 are 1.28, 2.43, and 1.2, respectively.

$$\overline{c}_{p,mix} = \frac{(0.161 \times 1.28 \times 44 + 0.081 \times 2.43 \times 18 + 0.758 \times 1.2 \times 28)}{1000}$$
$$= 0.0381 \text{ kJ/mol-K}.$$

Now, T_P is evaluated as,

$$T_P = (3169)/(37.2 \times 0.0381) = 2236 \text{ K}.$$

This value is close to the typical value measured from the experiments. The procedure can still be simplified. Since nitrogen is the primary constituent in the products of combustion using air, the specific heat of the mixture can be approximated as the specific heat of nitrogen at a given temperature. For instance, c_p of N_2 at 1200 K is 1.2 kJ/kg.K or 1.2 × 28/1000 = 0.0336 kJ/mol.K. Using this value,

$$T_P = (3169)/(37.2 \times 0.0336) = 2535 \text{ K}.$$

It is clear that the estimation of flame (product) temperature can be done by several procedures with various assumptions. Based on the application and necessity of the accuracy of the data, a proper approach with necessary details is adopted. In many cases, basically the interest is to determine the maximum possible temperature reached due to a fire hazard. For such calculation, common engineering assumptions, as illustrated above, can be invoked.

Analysis of Systems Till this point, a flow reactor (control volume) has been considered in the analysis of heat and temperature calculations. A system is considered to understand a device or closed compartment, where the mass remains constant during the analysis and volume may vary. As the first example, a piston-cylinder arrangement is considered, where combustion of a fuel with an oxidizer takes place in the volume in the cylinder, enclosed by the piston. The first law of thermodynamics for the system, in the absence of changes in potential and kinetic energies, is written as

$$Q - W = E_2 - E_1 = U_2 - U_1. \tag{2.16}$$

Here, Q is the heat interaction, W is the displacement work originating due to movement of the piston causing a volume change and $(U_2 - U_1)$ is the change in the internal energy. When a combustion reaction takes place inside the cylinder, the piston rises up to maintain the pressure constant at the value of the initial reactant pressure. The displacement work for a volume change of $(V_2 - V_1)$, is calculated as $p \times (V_2 - V_1)$. The first law is now written as

$$Q - p \times (V_2 - V_1) = U_2 - U_1.$$

$$Q = U_2 - U_1 + p \times (V_2 - V_1) = H_2 - H_1 = \Delta H_R. \tag{2.17}$$

It may be noted that the definition of enthalpy, which is an extensive property, is $H = U + pV$. Similar to Eq. (2.7), the standard enthalpy of a species is calculated as the sum of the enthalpy of formation and the sensible enthalpy. It is clear from Eq. (2.17) that the heat of the reaction for the combustion occurring in a constant pressure vessel (system) is the difference between the enthalpy of the product mixture and the enthalpy of the reactant mixture, similar to that in Eq. (2.5). For exothermic reactions, ΔH_R is negative and heat is transferred from system to the surroundings. Also, if the system is adiabatic, the products will reach the maximum temperature, the adiabatic flame temperature, such that the enthalpy of the product mixture is equal to the enthalpy of the reactant mixture.

On the other hand, if the system is a rigid vessel such that its volume remains constant, similar to a closed compartment, where a fire is initiated noting that displacement work will be zero as $\Delta V = 0$, the first law is written as

$$Q = U_2 - U_1 = U_P - U_R. \tag{2.18}$$

The heat of reaction in this case is the difference between the internal energy of the product mixture and the internal energy of the reactant mixture. The internal energy is obtained using the values of enthalpy, for which comprehensive data is available in the literature for various species as a function of temperature and pressure. For example,

$$U_P = H_P - p_P \times V,\ U_R = H_R - p_R \times V. \tag{2.19}$$

In most of the applications related to fire in enclosures, taking into account of the volume and the maximum values of pressures, the gas mixture inside the system can be treated as an ideal gas mixture, which obeys the equation of state of the form: $pV = nR_uT$. Here, n is the number of moles of the mixture, and R_u is the universal gas constant (8.314 kJ/kmol-K). Using this, Eq. (2.19) is written as

$$U_P = H_P - n_P R_u T_P,\ U_R = H_R - n_R R_u T_R. \tag{2.20}$$

Here, n_P and n_R are number of moles of product and reactant mixtures, respectively, and T_P and T_R are their respective temperatures. The enthalpies are estimated at given temperatures using the usual procedure, since for ideal gases the pressure will not influence enthalpy. The adiabatic flame temperature in the case of constant volume combustion is estimated by equating the internal energy of the product mixture to the internal energy of the reactant mixture.

This section is concluded with an illustration of constant volume combustion through an example. Let a closed rigid container contain propene, C_3H_6, and 150% theoretical air at 100 kPa and 298 K. Let the mixture be ignited and complete combustion takes place. Let heat be transferred so that the temperature of the products is maintained at 1200 K. The final pressure and heat interaction per mol of fuel are determined as follows. The reaction is written as

$$C_3H_6 + 6.75(O_2 + 3.76N_2) \rightarrow 3CO_2 + 3H_2O + 2.25O_2 + 25.38N_2.$$

The final pressure is estimated using the equation of state, noting that the volume remains constant. State 1 is that of reactant and state 2 is that of products. The number of moles of reactants and products being $n_1 = 33.13$ and $n_2 = 33.63$, respectively, the final pressure is estimated using

$$\frac{p_2}{p_1} = n_2T_2/(n_1T_1) => p_2 = \frac{100(33.63 \times 1200)}{(33.13 \times 298)} = 408.76 \text{ kPa}.$$

Applying first law to the system,

$$Q = U_2 - U_1 = (H_2 - n_2R_uT_2) - (H_1 - n_1R_uT_1).$$

The values of enthalpies are calculated by the usual procedure.

$$\begin{aligned} H_1 &= 1 \times \overline{h}_{(f,propene)} = 20.41 \text{ kJ} \\ H_2 &= 3(-393.52 + \Delta\overline{h}_{f,CO_2}(1200)) + 3(-241.83 + \Delta\overline{h}_{f,H_2O}(1200)) \\ &\quad + 2.25\Delta\overline{h}_{f,O_2}(1200) + 25.38\Delta\overline{h}_{f,N_2}(1200). \end{aligned}$$

Substituting the values from NIST database,

$$H_2 = 3(-393.52 + 44.47) + 3(-241.83 + 34.51) + 2.25 \times 29.7 + 25.38 \times 28.15$$

$$H_2 = -887.84 \text{ kJ}.$$

The heat interaction is calculated as,

$$Q = -887.84 - 33.63 \times 8.31415 \times 10^{-3} \times 1200 - (20.41 - 33.13 \times 8.31415 \times 10^{-3} \times 298) = -1161.69 \text{ kJ/mol-}C_3H_6.$$

2.4 Notes on Transport Processes in Fire

Fires involve strongly coupled transport of mass (multi-component mixture), momentum, species, and energy. The analysis of smoke transport is important in studying fire propagation, heat release rate, and the overall fire hazard. The basic physical aspects of the transport processes are briefly presented in this section, without presenting the governing equations and other details, which are available in Refs. [3, 4]. A governing equation, written as a partial differential equation, has four terms in general. They are the storage or the transient term, convective term, diffusion term, and the volumetric source term.

The mass transport involves conservation of mass of the mixture in a given control volume as a function of time. This has storage and convective terms. From the study of stoichiometry, it is found that mass is always conserved in a chemical reaction. That is, the mass of the reactants is equal to the mass of the products. Therefore, the mixture mass has to be conserved in a control volume, based on the constituent species present in the control volume at any time instant. Excluding explosions or fires in rigid close vessels, the pressure rise can be ignored in several fire scenarios. Because of the high temperature in fires, the mixture can be treated as an ideal gas mixture, obeying an equation of state given as $p = \rho RT$. Here, T is the temperature and p is the pressure (close to the atmospheric pressure in several applications). The quantity, R is the specific gas constant of the mixture (the product mixture is considered for smoke), calculated as universal gas constant divided by the molecular mass of the mixture (R_u/MW_{mix}). The molecular mass of the mixture is in turn calculated using mole fractions through Kay's rule; $MW_{mix} = \Sigma(\chi_i \times MW_i)$. The density of the mixture is calculated as $p/(RT)$ and is equal to the sum of the densities of the components forming the mixture. The mixture is transported at a velocity, which is measured using a probe or using laser-based techniques such as Particle Image Velocimetry (PIV). It should be noted that the density changes occurring in the flow field are due to the changes in the constituents forming the mixture and due to temperature changes. It is normally not affected by pressure unless the pressure rise is high. Thus, the mass of the mixture is conserved using parameters such as mixture density and mixture velocity. When two phases (condensed phase such as liquid or solid and gas phase) interact in a fire, then mass exchange occurs between these two phases. For example, a solid or liquid fuel gasifies and fuel gas/vapor is transported into gas phase. This interface transport is accounted by proper source terms.

Momentum of the gas mixture is conserved in a control volume at every time instant. The rate of change of momentum of the mixture is equated to the forces such as pressure, viscous, and gravity forces. The dynamic viscosity of the mixture (μ) is calculated as a function of temperature and species forming the mixture. The main force dominating the momentum of fire flow is the buoyancy force induced by gravity. This arises due to density difference between the cold and hot fluids and acts in the direction against that of the gravity vector. Kinematic viscosity (ν) of the mixture, calculated as dynamic viscosity divided by the density of the mixture, controls the molecular momentum exchange through diffusion. Fires are turbulent in nature. Even if the fire source is small, the buoyant plume arising from the fire may be in the turbulent regime. The molecular kinematic viscosity is enhanced several times by turbulent eddy interaction, and as a result, the momentum diffusion is enhanced significantly.

Species conservation is the next aspect to be carried out in the control volume. Species are transported by convection and diffusion. This is stated mathematically by Fick's law. Here the mass flux (kg/m^2-s) of a species, A, is written as

$$\rho_A v_A = \rho_A V + [D_{A,mix}(-\nabla \rho_A)]. \tag{2.21}$$

Here, v_A and ρ_A are the velocity vector and density of the species A, V is the mixture velocity vector, $D_{A,mix}$ is the diffusion coefficient of the species A diffusing into the mixture, $\nabla \rho_A$ is the gradient of density of species A that represents its mass fraction, $Y_A = \rho_A/\rho$ or concentration. The negative sign indicates that the diffusion of the species occurs in the direction of its decreasing concentration, similar to heat conduction that occurs in the direction of decreasing temperature. Fick's law states that the net mass flux of the species is the sum of its convective and diffusive fluxes. It should be noted that only *ordinary diffusion*, caused by concentration gradient, has been taken into account for species diffusion. In some cases, species diffusion can also occur because of temperature gradient. This is called thermal diffusion or the *Soret effect*. Equation (2.21) is written in term of species diffusion velocity, $V_{A,diff,X}$, as follows:

$$\rho_A v_A = \rho_A V + \rho_A V_{A,diff,X}. \tag{2.22}$$

The species diffusion velocity vector, $V_{A,diff,X}$ due to ordinary diffusion is calculated in terms of the mass fraction gradient and diffusion coefficient, given as,

$$V_{A,diff,X} = -D_{A,mix} \nabla Y_A / Y_A. \tag{2.23}$$

The species diffusion velocity is evaluated using set of differential equations called Stefan–Maxwell equations. This is a comprehensive approach. The expression shown in Eq. (2.23) is a simplified approach, where an algebraic expression dependent on mass fraction and binary diffusivities of species is used to calculate $D_{A,mix}$. The species diffusion as a result of temperature gradient, $V_{A,diff,T}$, is calculated in terms of thermal diffusion coefficient, $D_{A,T}$. The species diffusion velocity due to thermal diffusion is added to the diffusion velocity calculated for ordinary diffusion, written as

$$V_{A,diff} = V_{A,diff,X} + V_{A,diff,T}. \tag{2.24}$$

It should be noted that the mixture velocity, V, is measurable or predictable using a numerical model. The diffusion velocity, $V_{A,diff}$, is calculated as in Eqs. (2.23) and (2.24). The species velocity, v_A, is calculated as a vector sum of these two velocity vectors. The mixture velocity is also called mass averaged velocity, represented as $\Sigma(\rho_i v_i)/\rho$. Reactant species such as fuel and oxygen are consumed in fire and product species such as CO, CO_2, and H_2O are formed. Therefore, for each species, there is a net production (or consumption) rate. On mass basis, this has a unit of kg/m^3-s. This term is evaluated using a chemical kinetic mechanism and is written as a volumetric source term for species conservation.

Energy conservation equation includes a term for energy storage in the control volume (time dependent). Specific heat at constant pressure of the mixture is used to evaluate this term. The value of mixture specific heat is obtained by using specific heats of individual species forming the mixture, evaluated at the given temperature. The thermal energy transported by convection is evaluated using the difference between the net outgoing and net incoming convective energy fluxes. The energy transported by diffusion includes the heat transport by conduction in the mixture. Mixture thermal conductivity, calculated as a function of temperature and species mass fractions, is used in this term.

Diffusional transport of thermal energy also includes the sum of sensible enthalpy transported by the individual species at the given temperature. The diffusional flux of each species, calculated from the diffusion velocity of the species, is multiplied by the specific heat of the species and the temperature. The source term of the energy equation is the energy released from fire. This is calculated in a simple way by multiplying the net rate of consumption of fuel by the calorific value of the fuel. The heat loss due to radiation is also included as a volumetric sink (negative source) term. This includes radiation by gas species such as methane, carbon dioxide, carbon-monoxide, and water vapor. Radiation by soot is also considered to account for heat loss to the surroundings, such that realistic temperature prediction is obtained. For this, soot volume fraction and soot oxidation rates are estimated by several methods. In required cases, a complete radiation model, called discrete ordinate model is also used. In addition, a turbulence model is also required.

The mass, momentum, species, and energy transport processes are highly coupled. Any process is seen to affect the other processes. As a reactant is consumed and a product is formed, the composition of the mixture and its property change. Temperature increases as the result of heat release rate. This in turn bring in changes to the properties such as density. The buoyancy force is affected due to change in density, and as a result, the source term of momentum equation varies. The flow field is modified, affecting the convective transport of mass, momentum, species, and energy. Temperature gradients also modify the diffusion velocity of the species, besides contributing to changes in the values of diffusion coefficients. In a turbulent regime, the diffusion of mass, momentum, and energy are enhanced significantly due to interaction of eddies of different scales. The mass, momentum, and thermal diffusion coefficients, D, ν, and α, respectively, do not remain as molecular level fluid properties anymore; they become flow properties. The buoyant flow is seen to have large-scale eddies (recirculation zones) and the reaction zone has significant fluctuations. The reaction rates are also affected by the interaction of eddies. Local extinction of flame and reignition at some other location is observed.

The interaction between heat and mass transport processes is important in fires of condensed fuel surfaces. The heat transfer from the hot gas phase to the surface of the liquid or solid fuel results in mass transfer, that is, release of gaseous fuel components from the condensed fuel surface to the gas phase. The heat flux to the surface is due to conduction and radiation heat transfer. The dominance of radiation increases as the fuel surface area increases. The thermal conductivity of fluid close to the interface and the absorptivity of the surface are the important properties associated with the heat transfer.

The mass transfer from the surface of liquid fuel depends upon the surface temperature and the associated vapor pressure. The liquid fuel surface is heated up to a temperature closer to its boiling point. The mole fraction of the vapor at the interface increases with the surface temperature. The gradient of vapor concentration from the interface towards the gas phase dictates the mass transfer rate. The fuel flow rate from the surface obeys Fick's law, that is, the fuel is transported due to convection as well as diffusion. In liquids, under steady burning regime, the heat from the gas phase to the interface is used to vaporize the liquid fuel. In this scenario, the vaporization rate is enhanced by clearing the fuel vapor from the interface by gas-phase convection.

Sublimation, or melting followed by evaporation, of non-charring solid fuels is similar to the vaporization of a liquid fuel. Thermodynamic equilibrium prevails under several scenarios. The gasification or pyrolysis of a charring type solid fuel is quite different than the evaporation of a liquid fuel. As per the temperature of the solid surface, the volatile gases are released at a rate controlled by a typical Arrhenius rate equation that is strongly dependent on temperature. The transport of fuel gases towards the gas phase follows Fick's law.

It is clear that the transport phenomena in turbulent reacting flow scenarios observed in fires are quite complex. As a result of this, modeling of fire has been done utilizing empirical (experimentally determined) correlations. Simple approaches such as zone models have been reported in literature. Experimental research of large-scale fires and comprehensive numerical modeling of the same are still not resolved completely.

Review Questions

1. What is stoichiometry?
2. What is conserved in an overall single-step chemical reaction?
3. What is meant by atom balance?
4. Define stoichiometric air-fuel ratio.
5. Evaluate the mass of oxygen required for complete combustion of a mixture of *n*-butane and propene present in a mass ratio of 1:1.3.
6. Define equivalence ratio and excess air.
7. A fuel with chemical formula $C_xH_yO_z$ undergoes combustion in 145% theoretical air. The product composition is 15.72% O_2, 5.27% CO_2, 5.17% H_2O, and 0.67% CO by volume with the N_2 occupying rest of the volume. Determine the chemical formula of the fuel.
8. What is standard enthalpy of a species?
9. Define enthalpy of formation.
10. Write the first law for control volume considering steady flow of reactants into it and products out of it.
11. Define heat of reaction, heat of combustion, and calorific value.
12. Proximate analysis of coal yields (on mass basis) 28.7% volatiles, 5.4% moisture, 29.5% fixed carbon, and 36.4% ash. The elemental composition of coal when subjected to ultimate analysis (on dry ash free basis) is 43.7% C, 3.2% H, 0.9% N, 9.7% O, and 0.7% S. Calculate the stoichiometric air required for combustion per kg of coal (Nitrogen and Sulfur may be treated as inert). The volatiles are composed of CO_2, CO, H_2, and CH_4. Assuming the volatiles contain 20% CO_2 by volume, calculate the amount of air required for combustion of volatiles per kg of coal. Also calculate the standard heat of combustion of volatiles.
13. Why is heat of reaction negative in a combustion process?
14. What is the difference between higher and lower heating values?
15. The lower heating value of a liquid fuel of the form C_xH_{2x} is 45 MJ/kg-fuel. If the heat of formation of the fuel is −21 MJ/kmol, determine the higher heating value of the fuel.
16. Calculate the heat of combustion of *m*-Xylene when it exists in liquid and gaseous state. The standard enthalpy of formation and enthalpy of vaporization of *m*-Xylene are −25400 kJ/kmol and 35660 kJ/kmol, respectively.
17. What is the average value of heat of combustion per kg of oxygen consumed?
18. A methanol pool diluted with 5% water by mass is ignited and sustains a diffusion flame. Calculate the standard enthalpy of combustion per kg of fuel and per kg of O_2. The standard enthalpy of formation of methanol is taken as −250600 kJ/kmol.
19. How is adiabatic flame temperature calculated?
20. Char (carbon) at 900 °C ($h = 15267.17$ kJ/kmol) reacts with steam at 400 °C ($h = -228629.87$ kJ/kmol) to produce CO and H_2 at 800 K. Calculate the heat required for the reaction per kg of carbon and per kg of coal (consider the composition of coal in problem 12), the stoichiometric air required for burning

the gases produced (per kg of coal), the heat of combustion of gases (per kg of coal) when products are cooled to 298 K, and the adiabatic flame temperature.

21. A rigid insulated closed container has butane at 1 atm and 298 K. Air is supplied at 1 atm, 700 K such that a lean mixture ($\Phi = 0.9$) is formed. Ignition occurs and products (CO_2, H_2O, O_2, and N_2) are formed. Calculate the heat of combustion, the adiabatic flame temperature, and the final pressure. The standard enthalpy of formation of butane is taken as -125600 kJ/kmol.
22. What is equilibrium flame temperature?
23. Write first law for constant volume reactor.
24. Evaluate the adiabatic flame temperature of butane-air mixture in a rigid vessel and compare its value considering products to have nitrogen in abundance.
25. Calculate the adiabatic flame temperature at constant pressure when methane burns in oxygen at an equivalence ratio of 0.8.
26. Write Fick's law using an expression.
27. What is Soret effect?
28. How is species velocity calculated?
29. A binary mixture of A and B has a molecular weight of 30 kg/kmol and is present in a system at 1 atm and 300 K. Mass fraction of A varies as $Y_A(z) = 0.8 - 0.2z - 6.25z^2$ with distance z. If the binary diffusivity is 3.5×10^{-5} m^2/s and the bulk velocity of 0.03 m/s, evaluate net flux (in kg/m^{2}s) of B at $z = 0.2$.
30. What are the processes contributing to energy diffusion?
31. What is the effect of turbulence-kinetics interaction?

References

1. Turns, S.R. 2011. *An Introduction to Combustion - Concepts and Applications*, 3rd ed. New York: McGraw-Hill Publishers.
2. Rangwala, A.S., V. Raghavan, J. Sipe, and T. Okano. 2009. A New Property Evaluation Scheme for Mass Transfer Analysis in Fire Problems. *Fire Safety Journal* 44 (4): 652–658.
3. Faghri, M. and B. Sunden. 2008. *Transport Phenomena in Fires*. WIT Press.
4. Yeoh, G. H. and K. K. Yuen. 2008. *Computational Fluid Dynamics in Fire Engineering-Theory, Modelling and Practice*. Oxford: Butterworth-Heineman.

Chapter 3
Diffusion and Premixed Flames Related to Fires

Fires involve reactants, usually fuel and air, not intimately mixed at a molecular level before combustion. Usually, the fuel is in the solid or liquid state and thus transfer of material across a phase boundary (phase change) must also occur. The vaporized fuel must combine with oxygen from air to form a flammable mixture, which when ignited forms the flame zone. In most fire problems, this mixing of fuel vapor and oxygen takes place mostly by diffusion and takes orders of magnitude longer time compared with that of a chemical reaction. Therefore, diffusion of species is the primary controlling process during such burning behavior. A fundamental understanding of diffusion flames then involves exploring the mechanisms associated with the transport of the reactants and the resulting flame structure.

In some cases like compartment fire flash over (discussed in Chap. 7), deflagrations, and fire balls, the fuel vapor and oxygen are mixed before the burn. Such flames are called premixed flames and form the basis of explosion safety.

In both premixed and non-premixed flames the fundamental question is what is a flame and why does it arise? In both cases, the most striking feature of a flame is the steepness of the temperature. In addition to the steep temperature gradient, concentration gradients are also large in the flame zone where a strong exothermic reaction takes place in an extremely small volume. As will be explained in this chapter, the main difference is that in a diffusion flame, the fuel and oxidizer approach the reaction front from opposite sides. Thus the movement of a diffusion flame is

A. S. Rangwala and V. Raghavan, *Mechanism of Fires*,
https://doi.org/10.1007/978-3-030-75498-3_3

caused in the direction of the limiting reactant, which is usually fuel. This can be visualized by an example of a flame propagating on a matchstick.

In a premixed flame, the reaction zone separates the burned fuel from the unburned fuel. Thus the flame will move in the direction where there is an unburned fuel-air mixture and its propagation speed will be a function of the capability of the flame to consume the unburned reactants, which depends on the reaction rate.

In Fire Protection Engineering, diffusion flame theory is used in calculating flame length, flame location, and rates of burning. The flame length is used for hazard analysis as it provides information to estimate the heat transfer to surrounding surfaces. Knowledge of flame location is necessary for suppression, and finally, the rate of burning provides an estimate of the size of the fire and in combination with the heat of combustion is used to calculate the heat release rate. Similarly, premixed flame theory is used to calculate the burning velocity or rate of fuel-air premixture consumption. The burning velocity is used to calculate the pressure rise as a function of time in a compartment, ultimately used in the design of vents for explosion safety. Additionally, the burning velocity is also used to explain the limits of flammability, ignition, and quenching, which are important elements in the design of explosion prevention systems. It should be noted that while the time scales of influence of a diffusion flame heating in a compartment are of the order of minutes, the time scale of pressure rise because of the propagation of a premixed flame in a closed compartment is of the order of milliseconds. This difference in time scale is the reason for different solution strategies for fire and explosion hazards in industry. Both diffusion and premixed flames will be discussed in this chapter with a focus on application to fire safety. We will begin with diffusion flames which is adapted from the SFPE Handbook Chapter by one of the co-authors [1].

3.1 The Diffusion Coefficient

Diffusion is the phenomena of migration of mass. The mass can be in the form of atoms, molecules, ions, or other particles because of spatial gradient of some quantity (concentration, temperature, pressure, etc.) Similar to conduction heat transfer (Fourier's law) and momentum transfer (Newton's law), mass transfer is governed by a law called as Fick's law of diffusion. In a simplified context, Fick's law of diffusion describes the movement of one chemical species A through a binary mixture of A and B because of concentration gradient of A. In most fire problems A is usually fuel vapor, oxidizer, or products of combustion, while B represents air. To explain this further, let us consider the example of a candle flame shown in Fig. 3.1. The paraffin of the candle melts because of heat from the flame; it travels by capillary forces through the wick where it then evaporates to become paraffin vapor, a gaseous fuel.

Let F represent fuel vapor, O represent oxygen and P represent products. Fuel vapor will issue out of the wick because of the heat received from the flame. If one traverses along the path $X - X'$ (shown by dashed red line in Fig. 3.1), the concentra-

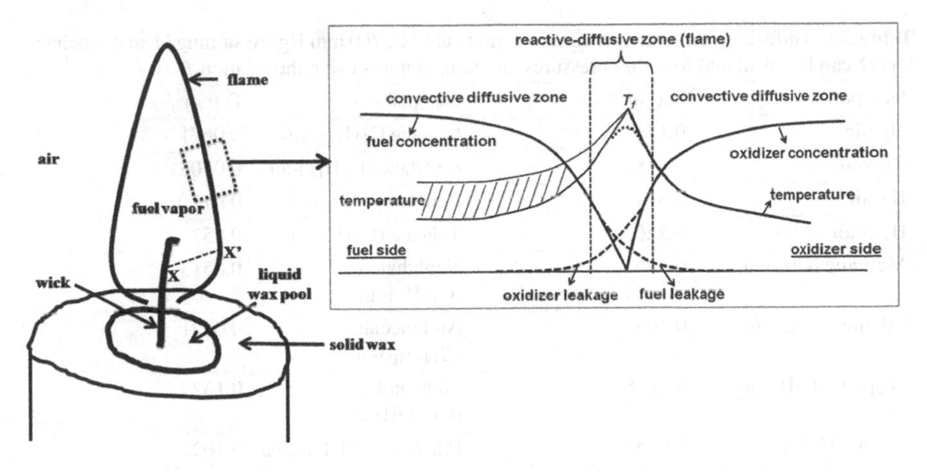

Fig. 3.1 The candle flame [1]

tion of fuel vapor is highest at the wick and reduces until it reaches a concentration most suitable for chemical reaction with oxygen at X'. The concentrations of the various species involved can be given by the mass fractions, Y_F, Y_O, and Y_P, where Y_F denotes the mass of fuel vapor divided by the total mass of gas mixture in a given volume and the subscripts O and P denote oxidizer and products, respectively.

Initially the mass fraction of F at the wick is 1 and zero just outside the wick, and this gradient drives the fuel vapor out of the wick. On approaching the flame zone, the mass fractions of the fuel and oxidizer should be such that a stoichiometric mixture should be formed. Now, Fick's law states that the mass transport of fuel vapor along XX' due to mass diffusion can be described by:

$$J_F = \rho D_{FA} \frac{Y_{F,X} - Y_{F,X'}}{XX'}. \tag{3.1}$$

In Eq. (3.1), J_F is the mass flow rate of fuel vapor per unit area (or mass flux) and is proportional to the mass fraction difference divided by the distance from the wick to the flame (XX'). ρ is the density of the gas-mixture system and D_{FA} is a proportionality factor called as the binary mass diffusivity of fuel vapor with respect to air. In differential form, Eq. (3.1) can be written as:

$$J_F = -\rho D_{FA} \frac{dY_F}{dx}. \tag{3.2}$$

The negative sign denotes that the mass fraction of fuel vapor will decrease as one moves along XX', represented as x-direction. This is logical because the fuel vapor originates at the wick (location X). Equation (3.2) is also called Fick's law of diffusion and forms the starting point of our discussion on diffusion flames. Note that similar

Table 3.1 Diffusion coefficients of common gases at 0 °C, 760 mm Hg. Assuming ideal gas behavior, D can be calculated for other pressures and temperatures using the relation $D \propto P^{-1}T^{3/2}$

Gas-pair	D (cm^2/s)	Gas-pair	D (cm^2/s)
O_2-air	0.178	Pentane $C_{15}H_{12}$-air	0.0671
CO_2-air	0.138	*n*-Octane (C_8H_{18})-air	0.0505
H_2-air	0.611	Benzene (C_6H_6)-air	0.077
H_2O-air	0.220	Toluene (C_7H_8)-air	0.051
Methane (CH_4)-air	0.196	Naphthalene ($C_{10}H_8$)-air	0.0513
Ethane (C_2H_6)-air	0.108	Anthracene ($C_{14}H_{10}$)-air	0.0421
Propane (C_3H_8)-air	0.0878	Methanol (CH_3OH)-air	0.1325
Butane (C_4H_{10})-air	0.075	Ethanol (C_2H_5OH)-air	0.102

relationships can be written for oxygen (diffusing towards the flame) and products of combustion (diffusing on either side of the flame).

Table 3.1 gives the binary diffusion coefficients D for many common gases. Values refer to atmospheric pressure. The first part of the table gives data for several gases in the presence of a large excess of air. The mass diffusivity D is an important transport property such as thermal diffusivity α and momentum diffusivity (kinematic viscosity) ν, and all three have dimensions of (length)^2/time. The ratios of these quantities, taken as a pair, form three important non-dimensional numbers that play a prominent role in analyzing most fire problems. They are Prandtl number, $\mathrm{Pr} = \nu/\alpha$, Schmidt number, $\mathrm{Sc} = \nu/D$ and Lewis number, $\mathrm{Le} = \alpha/D$. Equation (3.2) assumes that the driving force for diffusion is the concentration gradient. As discussed in Chap. 2, mass diffusion can occur in the presence of a temperature gradient and pressure gradient. Further, the value of the binary mass diffusivity itself depends on temperature and pressure ($D \sim P^{-1}T^{3/2}$). The diffusion coefficient of common gas mixtures at various temperatures and pressures can be obtained using this equation in combination with values listed in Table 3.1. However, this effect is usually neglected in fire problems. Further details on the subject can be found in Arias-Zugasti and Rosner [2].

3.2 Structure of Diffusion Flame

The zoomed inset of the part of the flame zone in Fig. 3.1 shows an illustrative sketch of the structure of a diffusion flame. It consists of a flame separating a fuel-rich zone and an oxidizer-rich zone. The flame or reaction zone incorporates the location of the maximum temperature. For example, for a methane-air flame, this temperature is experimentally observed to be around 1950 K (Smyth et al. [3]). As shown in

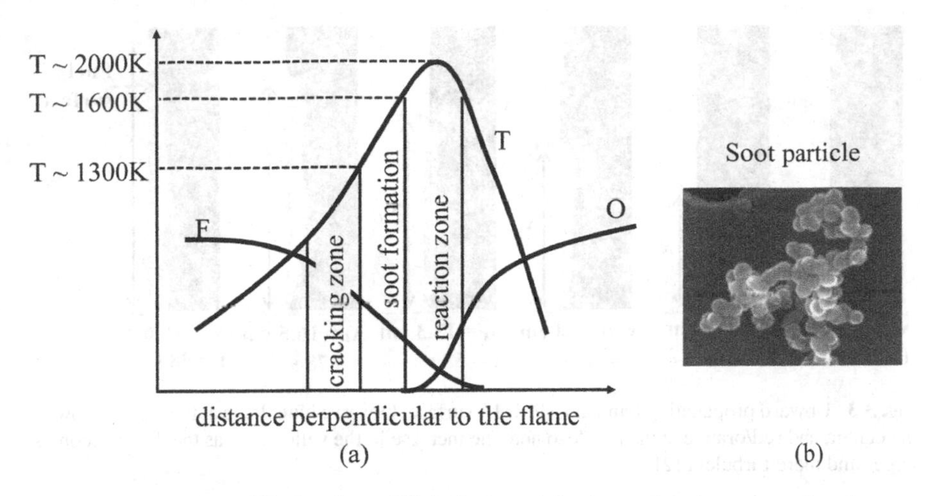

Fig. 3.2 **a** Zones in a diffusion flame [5]. **b** Soot particle observed under a scanning electron microscope—resemblance to string of pearls [6]

Fig. 3.1, fuel and oxidizer both almost disappear in the flame zone, although there is some fuel and oxidizer leakage outside the designated flame zone as shown by the dashed lines. Products of combustion and heat diffuse outwards from the flame zone to both the sides. One of the characteristics of typical hydrocarbon diffusion flames is their yellowness, especially when the fuel can emit soot, and its appearance can be explained using Fig. 3.2 [4].

The reaction zone usually has a blue emission, especially when the fuel and oxidizer have been mixed in proper proportions. This is mainly because of radiation due to excited CH radicals. The reddish glow arises from radiation from CO_2 and water vapor radiation. Most importantly, the intense yellow radiation which is a characteristic of most fires is due to the presence of carbon particles or soot. Figure 3.2a illustrates a simplified illustrative sketch of three prominent zones in a diffusion flame. Note that emphasis is given to the fuel side. The cracking zone is a region on the fuel side of the reaction zone before the soot formation zone. This is where the molecules crack and polymerize forming lighter fuel molecules which chemically react with the oxygen in the reaction zone and rest convert to carbonaceous and tarry substances in a soot formation zone that exists just before the reaction zone. Soot generally forms as particles with diameters of the order of several nanometers by a process called as inception or nucleation. These particles then undergo surface growth. One of the mechanisms attributed to the surface growth is called as the Hydrogen Abstraction by C_2H_2 Addition (HACA) mechanism [7] where H atoms impacting on the soot surface activate acetylene addition thereby increasing the mass of soot particles. The process of nucleation occurs concurrently with coagulation, where small particles coalesce to form larger primary particles, and agglomerations where multiple primary particles line up end-to-end to form larger structures resembling a string of pearls as shown in Fig. 3.2(b) [6]. When the soot particles pass through the flame

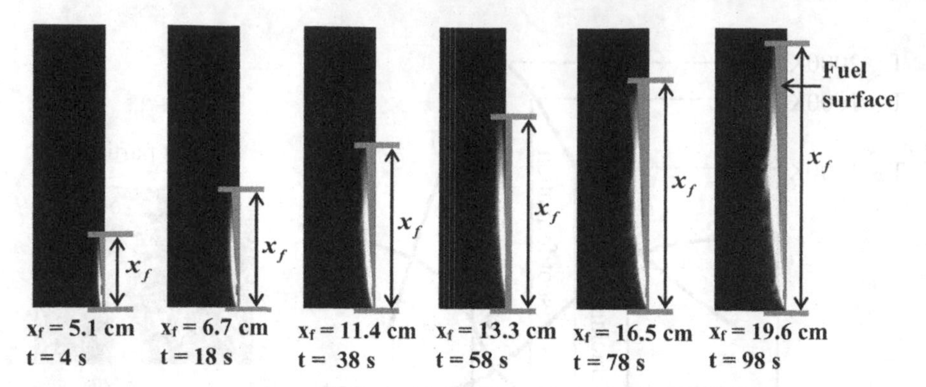

Fig. 3.3 Upward propagating flame on a PMMA surface. Observe blue flame at the base, yellow in the center, and red/orange at the tip. Also note the increase in the yellowness as the flame becomes larger and more turbulent [2]

front they oxidize whereby the mass of soot is decreased by heterogeneous surface reactions between soot particles and oxidizing species. Incandescent soot results in the characteristic yellow-color of most hydrocarbon diffusion flames. The processes leading to cracking and soot formation are dependent on fuel type and ambient conditions and form an important research topic in fire safety. This is mainly because most of the flame radiation from fires originates from soot particles. The luminous radiation from soot often makes it difficult to observe the blue emission from the reaction zone.

The blue radiation is observable for small flames (less than 15 cm) as shown in Fig. 3.3 where the initial stages of a wall fire (PMMA, 1.2 cm thick, 5 cm wide, 50 cm long) are shown. The size of the soot particles depends on the time allowed for their growth, the fuel composition, and temperature of the flame zone.

Since time increases with linear dimensions longer flames are usually yellower and smokier than shorter flames, indicating the escaping of the soot particles out of the flame zone. Large carbon particles will radiate more heat as shown in Fig. 3.3. A laminar natural convection flame is blue at the base, yellow in the center, and red/orange at the tip. The latter condition means that so much heat has been lost by radiation from the carbon particles, that when it does cross the reaction zone into a region where oxygen is available for combustion, it is too cold to burn. The result is emission of smoke. A useful material-property called smoke point of a fuel can be defined based on this radiative heat loss mechanism by soot. A fuel's smoke point is the maximum height of its laminar flame (or fuel mass flow rate) burning in air at which soot is just released from the flame tip. Another definition (and more applicable to fire safety) is the heat release rate at which smoke just begins to be released from the flame tip. Smoke point is a simplified ranking scheme for soot production and was first introduced by Kent and Wagner [7]. Smoke point can be easily determined for gases and vapors by adjusting the flow rate of the fuel from a simple burner. For liquid fuels a wick-fed lamp (ASTM D1322) is used. Determining the smoke point for solid fuels is difficult, although some progress has been made in this direction by de Ris and Cheng [8]. It has been shown [9, 10] that smoke point can provide a convenient measure of the flame radiant fraction. A comprehensive review on the development of an engineering model capable of predicting the release of soot and radiation given the smoke point of the fuel, stoichiometric mass ratio of the reactants, and the adiabatic stoichiometric flame temperature is discussed by Lautenberger [11].

3.3 Diffusion Flame Theory

The theory of diffusion flames consists of an analysis of factors controlling the mixing of fuel and oxidizer. The main factors controlling the mixing are mass diffusivity (D), gradient of species mass fraction (dY/dn) normal to the condensed fuel surface and the flow field. Unlike premixed flame analysis, the rates of the reaction mechanisms do not dominate the burning behavior in diffusion flames. As discussed earlier, in diffusion flames, fuel and oxidizer come together in a reaction zone through diffusion. This diffusion can be just molecular transport (candle flame, laboratory flames) or be enhanced several times by convection, which may be even turbulent (most large-scale fires such as pool, building, forest etc.) The theoretical solution of the diffusion flame is best approached by considering a candle flame once again. Focus on a control volume in the gas phase as shown by the dashed box in the zoomed inset in Fig. 3.1. The fuel vapor and oxidizer diffuse from opposite directions and approach the flame in a normal direction. The concentrations of fuel and oxidizer at the flame are in stoichiometric proportion. In other words, the diffusion flame surface is defined as the locus of all points in space where the fuel and oxygen meet at stoichiometric proportions. A one-step chemical reaction given by Eq. (3.3) can be used to represent the overall chemical process.

$$1\ g[\text{Fuel}] + sg[\text{Oxidizer}] \rightarrow (1+s)g[\text{Products}] + \text{Heat}. \tag{3.3}$$

The assumption made here is that the net disappearance rate of the reactants (fuel and oxidizer) is infinitely fast. This is represented by the zoomed inset in Fig. 3.1, where solid lines are used to indicate the profiles of temperature and mass fractions of the reactants. The flame zone is infinitesimally thin, and both fuel and oxygen are consumed at this zero-thickness flame sheet. However, in the actual scenario, the assumption of infinitely fast reaction is not true as indicated by the profiles in dashed lines in the flame structure inset in Fig. 3.1, where the flame zone has finite thickness and both the oxygen and the fuel leak through this flame zone. An interested reader may refer to a book related to the topic [12]. For our purposes, the infinite rate chemistry assumption is sufficient in predicting parameters such as the mean flame zone location and mass burning rate. The one-step and infinitely fast reaction assumptions also imply that

$$\dot{\omega}_F''' = \frac{\dot{\omega}_O'''}{s} = -\frac{\dot{\omega}_P'''}{1+s},$$

where $\dot{\omega}'''$ denotes the non-linear rate term representing the rate of formation or destruction of a species per unit volume. Subscripts F, O, and P, denote fuel, oxidizer, and products, respectively. The conservation equations for the control volume are given by:

—Species conservation (Assuming binary diffusion coefficients are equal for all species):

$$\rho\frac{\partial Y_O}{\partial t} + \rho u_i\frac{\partial Y_O}{\partial x_i} = \frac{\partial}{\partial x_i}\rho D\frac{\partial Y_O}{\partial x_i} - \dot{\omega}_O'''; \tag{3.4}$$

$$\rho\frac{\partial Y_F}{\partial t} + \rho u_i\frac{\partial Y_F}{\partial x_i} = \frac{\partial}{\partial x_i}\rho D\frac{\partial Y_F}{\partial x_i} - \dot{\omega}_F'''; \tag{3.5}$$

$$\rho\frac{\partial Y_P}{\partial t} + \rho u_i\frac{\partial Y_P}{\partial x_i} = \frac{\partial}{\partial x_i}\rho D\frac{\partial Y_P}{\partial x_i} + \dot{\omega}_P'''. \tag{3.6}$$

—Energy conservation:

$$\rho C_P\frac{\partial T}{\partial t} + \rho u_i C_P\frac{\partial T}{\partial x_i} = \frac{\partial}{\partial x_i}\lambda\frac{\partial T}{\partial x_i} + \dot{\omega}_F'''\Delta H_c \text{ or } \left[\frac{\dot{\omega}_O'''\Delta H_c}{s}\right] \text{ or } \left[\frac{\dot{\omega}_P'''\Delta H_c}{1+s}\right]. \tag{3.7}$$

In the above equations, ρ represents the gas phase density, C_p represents the specific heat and λ equals the thermal conductivity. ΔH_c represents the heat of combustion of the fuel and D equals the diffusion coefficient which is assumed to be the same for oxygen—air, fuel—air and product—air. The non-linear rate terms ($\dot{\omega}'''$) can be eliminated from the equations by suitable subtractions and assuming that the Lewis number is unity

$$Le = \frac{\lambda}{\rho C_P D} = 1.$$

Multiplying Eq. (3.4) by $\frac{1}{s}$, Eq. (3.6) by $\frac{1}{1+s}$ and Eq. (3.7) by $\frac{C_p}{\Delta H_c}$, the modified conservation equations are obtained.

—Modified species conservation equations:

$$\rho\frac{\partial(Y_O/s)}{\partial t} + \rho u_i\frac{\partial(Y_O/s)}{\partial x_i} = \frac{\partial}{\partial x_i}\rho D\frac{\partial(Y_O/s)}{\partial x_i} - \frac{\dot{\omega}_O'''}{s}; \tag{3.8}$$

$$\rho\frac{\partial Y_F}{\partial t} + \rho u_i\frac{\partial Y_F}{\partial x_i} = \frac{\partial}{\partial x_i}\rho D\frac{\partial Y_F}{\partial x_i} - \dot{\omega}_F'''; \tag{3.9}$$

Table 3.2 Different forms of the coupling function β

β	Value	β	Value
β_{FO}	$Y_F - \frac{Y_O}{s}$	β_{FP}	$Y_F + \frac{Y_P}{1+s}$
β_{FT}	$Y_F + \frac{TC_p}{\Delta H_c}$	β_{OP}	$Y_O + \frac{Y_P}{1+s}$
β_{OT}	$\frac{Y_O}{s} + \frac{TC_p}{\Delta H_c}$	β_{PT}	$\frac{Y_P}{1+s} - \frac{TC_p}{\Delta H_c}$

$$\rho\frac{\partial[Y_P/(1+s)]}{\partial t} + \rho u_i\frac{\partial[Y_P/(1+s)]}{\partial x_i} = \frac{\partial}{\partial x_i}\rho D\frac{\partial[Y_P/(1+s)]}{\partial x_i} + \frac{\dot{\omega}_P'''}{1+s}. \tag{3.10}$$

—Modified energy conservation equation:

$$\rho\frac{\partial(TC_p/\Delta H_c)}{\partial t} + \rho u_i\frac{\partial(TC_p/\Delta H_c)}{\partial x_i} = \frac{\partial}{\partial x_i}\frac{\lambda}{C_p}\frac{\partial(TC_p/\Delta H_c)}{\partial x_i} + \dot{\omega}_F'''. \tag{3.11}$$

Equations (3.8)–(3.11) can be combined into a single equation given by

$$L(\beta) = 0, \tag{3.12}$$

where β can take several values as shown in Table 3.2 and the operator L is expressed as,

$$L(\beta) \equiv \rho\frac{\partial\beta}{\partial t} + \rho u_i\frac{\partial\beta}{\partial x_i} - \frac{\partial}{\partial x_i}\rho D\frac{\partial\beta}{\partial x_i}. \tag{3.13}$$

In the operator L, the first term represents the accumulation of thermal energy or chemical species, the second term represents the convection efflux through the control surfaces and the third represents the diffusion efflux. The non-linear volumetric reactive effects are eliminated using the coupling function β, which can take six forms as shown in Table 3.2. This methodology is referred to as the Shvab-Zeldovich transformation after two classical papers by Shvab [13] and Zeldovich [14] that first used the coupling function. Note that although Shvab-Zeldovich proposed a general solution in 1950, the original idea was first proposed by Burke-Schumann [15] in 1928.

Equation (3.12) can be solved with knowledge of initial and boundary conditions. However, this is not an easy task! For example, the convective term is also non-linear unless the velocity is constant. Further, many added assumptions such as steady state $\left(\frac{\partial\beta}{\partial t} = 0\right)$, one-dimensional system, constant pressure, and low speed flow are required before analytical solutions to some problems can be obtained. Nevertheless the coupling function β is a powerful tool that is used extensively in diffusion flame problems.

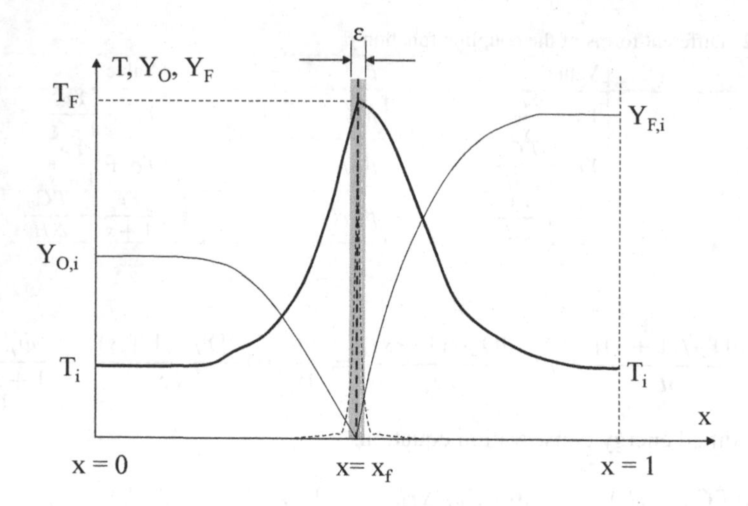

Fig. 3.4 The one-dimensional diffusion flame mathematical model [1]

3.4 Diffusion Flame Location

The diffusion flame surface is defined as the locus of all points in space where the fuel and oxygen meet at stoichiometric proportions. The position of the flame front of a diffusion flame is dependent on the surrounding geometry and the flow rates of the various gas streams. It is apparent, then, that each accidental flame is dependent on the details of the environment in which it burns. Thus the possible variations in behavior are endless. The one-dimensional flame structure discussed earlier will be used once again to describe the problem. A further set of assumptions will be imposed to simplify the math and facilitate physical understanding:

1. The oxidizer enters the system at $x = 0$ with a temperature T_i, a concentration of $Y_{O,i}$ and a velocity of $u = U$ which is a constant.
2. The fuel enters the system at $x = 1$ with a temperature T_i, a concentration of $Y_{F,i}$ and a velocity of $u = U$.
3. The reaction occurs at $x = x_f$ in a zone with thickness, $\epsilon \to 0$.
4. All reactants are consumed at the reaction zone so for $x < x_f$, $Y_F = 0$ and $x > x_f$, $Y_O = 0$.

Conservation Equations

Since the flow field is assumed to be known and constant ($u = U$), there is no need to solve the overall mass and momentum conservation equations.

Species Conservation

$$U\frac{d(Y_O/s)}{dx} = D\frac{d^2(Y_O/s)}{dx^2} - \frac{\dot{\omega}_F'''}{s\rho}; \tag{3.14}$$

$$U\frac{dY_F}{dx} = D\frac{d^2Y_F}{dx^2} - \frac{\dot{\omega}_F'''}{\rho}. \tag{3.15}$$

Energy conservation:

$$U\frac{d(TC_p/\Delta H_c)}{dx} = \alpha\frac{d^2(TC_P/\Delta H_c)}{dx^2} + \frac{\dot{\omega}_F'''}{\rho}, \tag{3.16}$$

where $\alpha = \frac{\lambda}{\rho C_p}$ represents the thermal diffusivity..

Boundary Conditions

$$\begin{aligned} x = 0 &: T = T_i,\ Y_O = Y_{O,i},\ Y_F = 0\,; \\ x = 1 &: T = T_i,\ Y_O = 0,\ Y_F = Y_{F,i}. \end{aligned} \tag{3.17}$$

The conservation equations are coupled using Eq. (3.12). Note that with the steady, one-dimensional system discussed in the example, the L operator reduces as shown.

$$L(\beta) \equiv \rho U\frac{d\beta}{dx} - \rho D\frac{d^2\beta}{dx^2}.$$

This gives,

$$U\frac{d\beta}{dx} = \alpha\left(\frac{d^2\beta}{dx^2}\right). \tag{3.18}$$

Three β variables will be selected such that the values of temperature and mass fractions of fuel and oxygen (three variables obtained by solving three equations) can be evaluated later. The choices are β_{OF}, β_{OT}, and β_{FT}. The variables and the boundary conditions are listed below:

$$\begin{aligned} &\beta_{OF} = \frac{Y_O}{s} - Y_F,\ \beta_{OT} = \frac{Y_O}{s} + \frac{TC_p}{\Delta H_c},\ \beta_{FT} = Y_F + \frac{TC_p}{\Delta H_c}; \\ x = 0:\ &\beta_{OF} = \frac{Y_{O,i}}{s},\ \beta_{OT} = \frac{Y_{O,i}}{s} + \frac{T_iC_p}{\Delta H_c},\ \beta_{FT} = \frac{T_iC_p}{\Delta H_c}; \\ x = 1:\ &\beta_{OF} = -Y_{F,i},\ \beta_{OT} = \frac{T_iC_p}{\Delta H_c},\ \beta_{FT} = Y_{F,i} + \frac{T_iC_p}{\Delta H_c}. \end{aligned} \tag{3.19}$$

As shown above, the coupling function or the Shvab-Zedovich transformation eliminates the non-linear reaction rate terms from the conservation equations. We are still left with 6 boundary conditions for the three linear equations that can be obtained with β_{OF}, β_{OT}, and β_{FT}. We can further decrease the difficulty in the solution procedure without altering the nature of the solution by replacing with a new normalized variable, $Z = \frac{\beta(x)-\beta(0)}{\beta(1)-\beta(0)}$, which is called as a mixture fraction. Note that $\beta(0)$ and $\beta(1)$ are constants obtained from the boundary conditions. Since the boundary conditions for the variable Z at the boundaries $x = 0$ and $x = 1$ are $Z = 0$ and $Z = 1$, respectively, irrespective of the β-variable that defines Z, the solution is unique and we need to obtain the solution to one mixture fraction equation only. This simple algebraic manipulation allows us to obtain a single differential equation with normalized boundary conditions given by

$$U\frac{dZ}{dx} = \alpha\left(\frac{d^2Z}{dx^2}\right), \text{ where } x = 0 : Z = 0, \ x = 1 : Z = 1. \tag{3.20}$$

The solution to this differential equation is

$$Z = \frac{(e^{x/\delta} - 1)}{(e^{1/\delta} - 1)}, \text{ where } \delta = \frac{\alpha}{U}. \tag{3.21}$$

Going back to the dimensional values (oxygen concentration, fuel concentration and temperature) requires the determination of the flame location. Based on the assumption that the flame will place itself where fuel and oxidizer arrive in stoichiometric proportions, and at the flame location, they will be consumed fully, the flame location can be expressed by $\beta_{OF}(x_f) = 0$. Substituting the appropriate expression for $Z(x)$ as

$$Z = \frac{\beta_{OF}(x) - \beta_{OF}(0)}{\beta_{OF}(1) - \beta_{OF}(0)}, \text{ where } \beta_{OF}(0) = Y_{O,i}/s, \text{ and } \beta_{OF}(1) = -Y_{F,i}$$

$$\beta_{OF}(x) = \frac{Y_{O,i}}{s} - \left(\frac{Y_{O,i}}{s} + Y_{F,i}\right)\left[\frac{(e^{x/\delta} - 1)}{(e^{1/\delta} - 1)}\right]. \tag{3.22}$$

At the flame location, $x = x_f$, Eq. 3.22 equates to zero, and the flame location, x_f, is obtained as

$$x_f = \delta \ln\left[\frac{Y_{O,i}}{s}\frac{(e^{1/\delta} - 1)}{\left(\frac{Y_{O,i}}{s} + Y_{F,i}\right)} + 1\right]. \tag{3.23}$$

Finally, assuming there is no oxidizer in the fuel zone and no fuel in the oxidizer zone the fuel and oxygen concentrations can be defined as:

$$x < x_f : Y_O = s\beta_{OF}(x) \text{ and } Y_F = 0 ;$$
$$x > x_f : Y_O = 0 \text{ and } Y_F = -\beta_{OF}(x). \tag{3.24}$$

A similar method can be followed when determining the temperature distribution from β_{FT} for $x < x_f$ and β_{OT} for $x > x_f$. There are several advantages in introducing the mixture fraction. The mixture fraction (Z) should satisfy the balance equation

$$L(Z) \equiv \rho\frac{\partial Z}{\partial t} + \rho u_i \frac{\partial Z}{\partial x_i} - \frac{\partial}{\partial x_i}\rho D \frac{\partial Z}{\partial x_i} = 0, \tag{3.25}$$

based on the definition of the operator L of Eq. (3.10). The boundary condition equation is $Z = 1$ in the fuel stream and $Z = 0$ in the oxidizer stream. This equation converts all the Shvab-Zeldovich variables (β) into a single parameter. Equation (3.25) greatly simplifies the modeling of diffusion flames. Note that there is no source term in Eq. (3.25).

It is important to reiterate the assumptions used in working out the solution discussed earlier:

1. One-step and infinitely fast reaction.
2. Lewis number = 1.
3. Binary diffusion coefficients are equal for all species.
4. ρD = constant.
5. Velocity (U) is constant.
6. Steady state.
7. One dimensional.
8. Constant specific heat, thermal diffusivity, and density.

Assumptions 1, 2, and 3 are necessary for implementing the Shvab-Zeldovich transformation. This is because Z couples the transport of heat and species into a single variable. In some cases, heat and different species may have different diffusivities and therefore can be transported at different rates. Consequently, β and Z are no longer conserved. However, by making the equal diffusivity and Le = 1 assumption, we are requiring that these diffusive fluxes are transported at the same rate and hence preserving the conserved nature of Z. Assumption 4 is used in most fire problems, however, can be relaxed. Assumptions 5–8 were used to solve the specific example of a one-dimensional flame. Of these 5 can be most questionable. This is because the flow field plays a significant role in diffusion problems.

An exact representation of the flow field can be obtained by solving the overall mass and momentum conservation or the Navier-Stokes equations, which require proper pressure-velocity coupling since most fires are incompressible in nature. Assumption 6 is reasonable mainly because of the slow regression rates observed for most condensed fuels. Assumption 7 facilitated an analytical solution. Assumption 8 is reasonable so long as the properties are chosen correctly. The correct choice of properties is hugely important in all problems of this nature. This is further discussed below.

Property Estimation Most theoretical and empirical expressions to solve fire problems usually rely on the assumption of constant thermo-physical properties. An important issue in using these expressions in practice therefore necessitates a proper method to evaluate the thermo-physical properties such that results obtained through them match with experimental data. The first step in property estimation is obtaining a certain average temperature.

For a diffusion flame, the temperature can vary from a few hundred degrees at the fuel side to around 1500–2000 K in the flame zone. The surrounding temperature can also be of the order of a few tens to a few hundred degrees based on the problem. Composition of the gas mixture within this range varies from a pure fuel vapor near the interface to pure air in the far-field. Within this range, there exists many gas species formed because of thermal cracking of the fuel vapor as well as combustion products such as CO_2, CO, and H_2O. Therefore the process of arriving at an average mixture property based on specific mixture composition and average temperature is a non-trivial issue.

This problem of an average gas-composition at an average temperature whose properties can be used for correlations with constant property assumptions has been investigated by several researchers (cf. Rangwala et al. [16] for a list of references related to the topic). The main method reported in most combustion textbooks is by Law and Williams [17] and employs flame, ambient (or surrounding) and interface temperature to arrive at an average temperature. The average mixture composition is calculated using some proportions of fuel and air. The disadvantage of using the Law and Williams [17] scheme in fire problems is the need to know the fuel vapor composition which is difficult to evaluate for complex materials usually involved.

A much simpler scheme using only properties of air was developed by Rangwala et al. [16]. The scheme has been tested in several diffusion controlled problems involving burning behavior of both gaseous liquid and solid fuels. In this scheme, the average thermal conductivity is estimated as the thermal conductivity of air calculated at a temperature given as one-third the sum of ambient and the adiabatic flame temperature. The gas phase specific heat is estimated as the specific heat of air at adiabatic flame temperature. Adiabatic flame temperatures for several fuels are tabulated in standard fire dynamics textbooks or can be calculated using computational tools discussed in Chap. 1.

3.5 Diffusion Flame Height

When gaseous fuel issues out of a tube into ambient atmosphere of air and the gas is ignited a flame is established as shown in Fig. 3.5. The question we will try to answer is how high is the resulting diffusion flame? The problem can be solved using conservation equations with certain approximations as was first shown by Burke and Schumann [15]. However, for practical purposes a simple physical reasoning exercise will be adopted here. It will be shown later that the relationship obtained is similar to that obtained using the more rigorous approach.

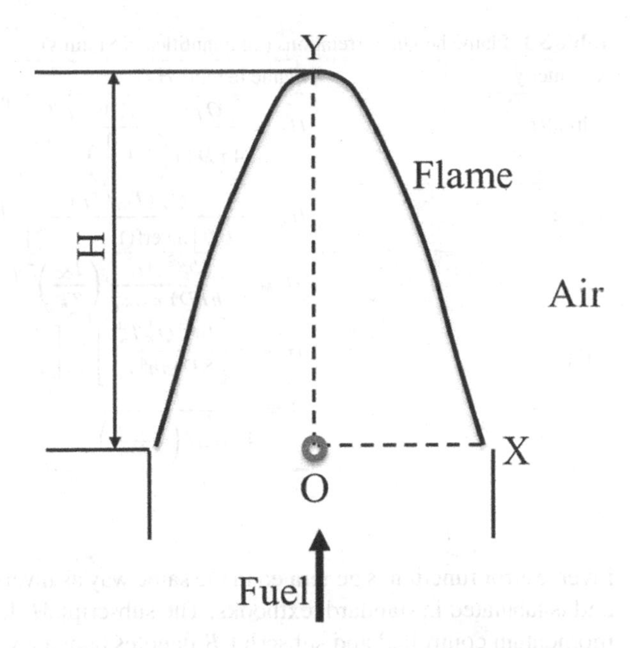

Fig. 3.5 The diffusion jet flame

Imagine a fuel molecule of diameter d, shown in Fig. 3.5 by the circle, initially located at the center of the burner tube. The molecule can traverse two extreme paths (depicted by O-X and O-Y) to meet with oxygen at the flame surface. The time taken to traverse horizontally along O-X is given by $d^2/4D$, where D is the diffusion coefficient between methane and air. The relationship is obtained by a dimensional analysis of the two parameters: distance over which diffusion must occur (m) and diffusion coefficient (m^2/s). The only way of obtaining a time scale from these two quantities is to divide the square of the distance by the diffusion coefficient. The length of the flame will correspond to the condition that a point on the stream axis where combustion is complete, the average depth of penetration of air into gas must equal to the radius of the burner tube.

Similarly, the time taken to traverse vertically along O-Y is given by H_f/V, where H_f is the flame height and V is the velocity of the gas issuing of the burner. Equating the two times gives, $H_f/V \sim d^2/4D$ or $H_f \sim d^2V/4D$. This simple result is reflected in all the correlations (developed by Roper [18, 19]) related to diffusion flame height shown in Table 3.3. An important element in the expression developed is the influence of fuel properties, geometry (of the duct). The influence of fuel properties is usually incorporated in the flame temperature and stoichiometric coefficient (s) shown in Table 3.3.

In Table 3.3, Q_F = volumetric fuel flow rate (m^3/s), D = diffusivity (m^2/s), T_∞ = ambient temperature (K), T_f = mean flame temperature (K), T_F = fuel temperature (K), s = molar stoichiometric oxidizer-fuel ratio, inverf = inverse error function, ω = inverf[erf(ω)], and a = mean buoyant acceleration $\approx 0.6g\left(\frac{T_f}{T_\infty} - 1\right)$ (m^2/s). The

Table 3.3 Flame height correlations (all quantities in SI units)

Geometry	Flame height (H_f)
Circular	$H_f = \dfrac{Q_F(T_\infty/T_F)}{4\pi D \ln\left(1+\dfrac{1}{s}\right)}\left(\dfrac{T_\infty}{T_f}\right)^{0.67}$
Square	$H_f = \dfrac{Q_F(T_\infty/T_F)}{16D\left[\text{inverf}(1+s)^{-0.5}\right]^2}\left(\dfrac{T_\infty}{T_f}\right)^{0.67}$
Slot	$H_{fM} = \dfrac{b\beta^2 Q_F}{hIDY_{F,stoic}}\left(\dfrac{T_\infty}{T_F}\right)^2\left(\dfrac{T_f}{T_\infty}\right)^{0.33}$ $H_{fB} = \left[\dfrac{9\beta^4 Q_F^4 T_\infty^4}{8D^2ah^4T_F^4}\right]^{1/3}\left[\dfrac{T_f}{T_\infty}\right]^{2/9}$ $\beta = \dfrac{1}{4\,\text{inverf}\left(1+\dfrac{1}{s}\right)}$

inverse error function is generated in the same way as inverse trigonometric functions and is tabulated in standard textbooks. The subscript M denotes that the flow field is momentum controlled and subscript B denotes buoyancy controlled.

To determine if the flame is momentum or buoyancy controlled the flame Froude number Fr must be calculated as

$$\text{Fr}_f = \frac{(V_e I Y_{F,stoic})}{aH_f}. \tag{3.35}$$

V_e is the velocity at the exit. The parameter, I typically takes values between 1 and 1.5. $I = 1$ for uniform flow and $I = 1.5$ for fully developed parabolic exit velocity profile. If $\text{Fr}_f >> 1$ then flame is momentum controlled and if $\text{Fr}_f << 1$ then the flame is buoyancy controlled. If $\text{Fr}_f \approx 1$, then the flame lies in a transitional zone which is both momentum and buoyancy controlled. In this case,

$$H_{fT} = \frac{4}{9}H_{fM}\left(\frac{H_{fB}}{H_{fM}}\right)^3\left(\left[1+3.38\left(\frac{H_{fB}}{H_{fM}}\right)^3\right]^{2/3}-1\right), \tag{3.36}$$

where H_{fM} and H_{fB} represent the momentum and buoyancy controlled flame heights and can be obtained using the correlation in Table 3.3. Note that for a circular and square port geometry the equations are applicable for both momentum and buoyancy controlled.

The equations shown in Table 3.3 demonstrate that some of the primary controlling parameters influencing the flame height of a diffusion flame are flow rate, geometry, and stoichiometry. The flow rate and geometry influence can be characterized by the non-dimensional Froude number. The molar stoichiometric ratio is mainly a function of fuel type. As discussed in Chap. 2, for a generic hydrocarbon C_xH_y burning in air the chemical balance equation assuming complete combustion can be written as

$$C_xH_y + a(O_2 + 3.76N_2) \rightarrow xCO_2 + \frac{y}{2}H_2O + 3.76aN_2. \qquad (3.37)$$

In Eq. (3.37), $a = x + \frac{y}{4}$ and

$$s = \left(\frac{\text{moles air}}{\text{moles fuel}}\right)_{stoic} = \frac{x + y/4}{X_{O_2}}.$$

X_{O_2} is the mole fraction of oxygen in air and equals 1/4.76. As the carbon content in the fuel increases (x/y increases) the flame length increases. For example, a propane flame (C_3H_8) is nearly 2.5 times longer than a methane flame (CH_4). Fuels which have oxygen in their molecule (e.g., CO) show an even smaller flame length. If the fuel is diluted with an inert gas, the flame height reduces. The molar stoichiometric ratio in this case is

$$s = \frac{x + y/4}{\left(\frac{1}{1 - X_{dil}}\right) X_{O_2}},$$

where X_{dil} is the mole fraction of the inert in the fuel stream.

The discussion so far pertains to gaseous jet flames. Roper's correlations can be extended to predicting flame height of condensed fuels as well. For example, an approximate fuel flow velocity in a candle flame can be obtained by assuming the candle wick to be a vertical cylinder. Using Roper's correlation for a circular burner (Table 3.3) and substituting $\dot{Q}_F = \frac{\dot{m}_F}{\rho_F}$ and using heat transfer coefficients for a finite vertical cylinder (represents candle wick) and a hot surface facing up (which represents the top face of candle wick), flame height can be evaluated as shown in Eq. (3.38) [20].

$$H_f \sim 0.5D\left(\frac{L}{D}\text{Ra}^{1/4} - 2\right)^n. \qquad (3.38)$$

In Eq. (3.38), D equals the diameter of the wick, L is equal to the length of the wick from the base, Ra is the Rayleigh number defined using the wick diameter, and n is a constant approximately equal to 0.7. Sunderland et al. [21] have shown the above relationship to provide a good match with experimental data using candle wicks of different thicknesses.

3.6 Turbulent Diffusion Flames

Most fires can be classified as turbulent diffusion flames, and extensive studies related to the basic understanding of the problem can be found in literature [22–25]. Besides the important aspects of height of a diffusion flame and rate of burning, which were discussed in the previous sections, an important fire safety concern with the advent of turbulence is the enhancement in the transport processes, changes in the flame structure, extent and shape, and the radiant energy flux emitted by the flame. In large fires, a solid fuel is normally gasified by radiative heat flux from the fire (and hot zones) and the evolved gases mix with surrounding air. The mixing process is determined by buoyancy induced flow rather than by the forced convection as in jets and sprays usually studied in the field of combustion in engines and furnaces, for example. The advent of turbulence, which is a random process, necessitates the use of statistical methods, thus making the solution probabilistic in nature. Thus an exact solution is seldom possible. Given our current state of understanding a solution that falls in a reasonable range is acceptable. The coupling function described as the mixture fraction (Eq. (3.25)) plays a central role in reducing problems in the analysis of turbulent diffusion flames to problems associated with non-reacting turbulent flows.

The turbulent flow field can be visualized as being comprised of eddies having wide range of length scales, the larger of which extract energy from the mean flow. The large eddies break up into smaller eddies, which in turn break up into even smaller ones, until the smallest eddies, called Kolmogorov eddies form, and dissipate the energy back to the main flow as viscous dissipation. This concept was first proposed by Kolmogorov is described as the eddy/energy cascade hypothesis in turbulent literature [23]. Note that the size of the smallest eddies is several orders of magnitude larger than the mean free path of the molecules of the gas or fluid. Due to this reason, the turbulent flow field can be considered as continuum and the conservation equations of mass (species), momentum, and energy are still applicable. The question then arises that how does this flow pattern influence a diffusion flame? For this, the general theory proposed by Williams [26] and models by Peters [25] and Kuznetsov [27] will be discussed. The general model is named as the flamelet model mainly because of the tendency of turbulence to break the flame into several tiny flames or flamelets. Both strain and shear at the interface of eddies increase the concentration gradient between the reactants thereby enhancing the mixing. Going back to Eq. (3.1) (Ficks law) as the gradient $\frac{\partial Y}{\partial x_i}$, increases, the rate of mixing should also increase. However, this increase is not due to a change in the molecular diffusivity D_{AB}. Thus, in order to model this increase in the overall mixing process, an improvised turbulent diffusivity is defined. Note that the turbulent diffusivity is an empirical concept and incorporates both the molecular diffusivity as well as some function which can represent the influence of intensity and scale of turbulence on the mixing at the flamelet level. The flamelets are considered as thin reactive-diffusive layers, considered in general as laminar, embedded within an otherwise non-reacting turbulent flow field. This allows us to use the mixture fraction to describe turbulent flames as well.

If we assume that the turbulent transport of momentum, species, and thermal energy are all equal (Sc = Pr = Le) we can substitute existing correlations from the fluid mechanics literature related to turbulent momentum diffusivity (eddy viscosity) to apply for turbulent mass diffusivity and thermal diffusivity as well.

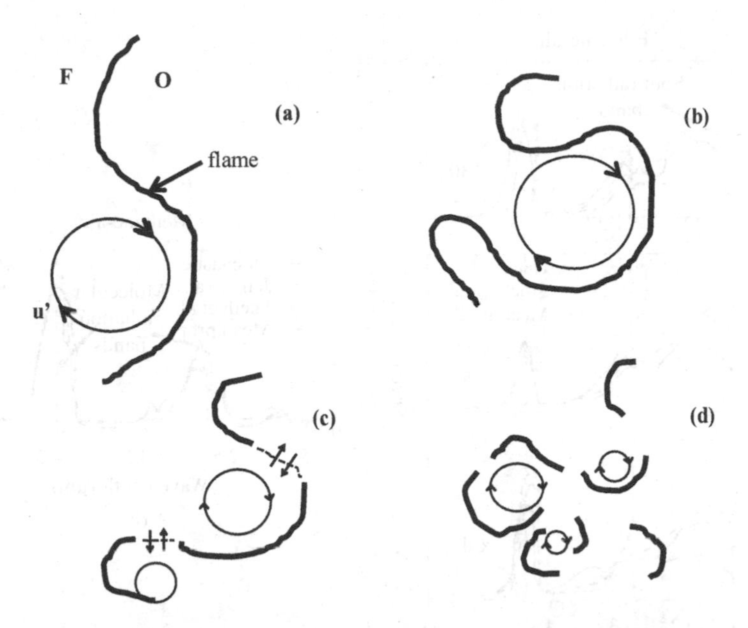

Fig. 3.6 Illustrative sketch of laminar flamelet concept of turbulent diffusion flames. The turbulent intensity increases from **a–d**. Local extinction is observed in **c** due to flame stretching [1]

Figure 3.6 shows the influence of increased turbulent intensity on a turbulent flame. As we move from Fig. 3.6a–d, eddies are capable of corrugating a flame and eventually breaking the corrugations and distributing the flame sheet into smaller flamelets. Thus at sufficiently high strain, there is also local extinction achieved. Another aspect, and most important from an aspect of fire safety is an understanding of flame radiation due to large turbulent hydrocarbon diffusion flames. For example, flame radiation is an important parameter in design of safe separation distances. The primary factors influencing the problem are the concentration of soot, CO_2, H_2O, temperature profiles (at different locations in the flame), and geometry of the problem. Incandescent soot formed within the flame is the primary source of flame radiation in large-scale fires, although gaseous products of combustion (CO_2, H_2O) can also emit heat. The radiation emitted from CO_2 and H_2O is usually called molecular radiation. The easiest way to understand (and explore) the difference between the radiation from soot and gaseous products (CO_2, H_2O) is by comparing two diffusion flames.

Figures 3.7a [28] and 3.7b [29] depict spectral radiation intensities for an ethylene-air and methane-air diffusion flame respectively. The ethylene-air flame produces more soot when compared with the methane-air flame. Correspondingly, it exhibits intensity peaks at 1.5 μm corresponding to the in-flame soot. In comparison, the methane-air flame shows significant intensity peaks at 2.5–3 μm and 4–5 μm corresponding to the molecular radiation from CO_2 and H_2O. Prediction of flame radiation necessitates knowledge of formation and oxidation of soot in the flame, its concentration profiles, as well as concentration and temperature of sources of molecular

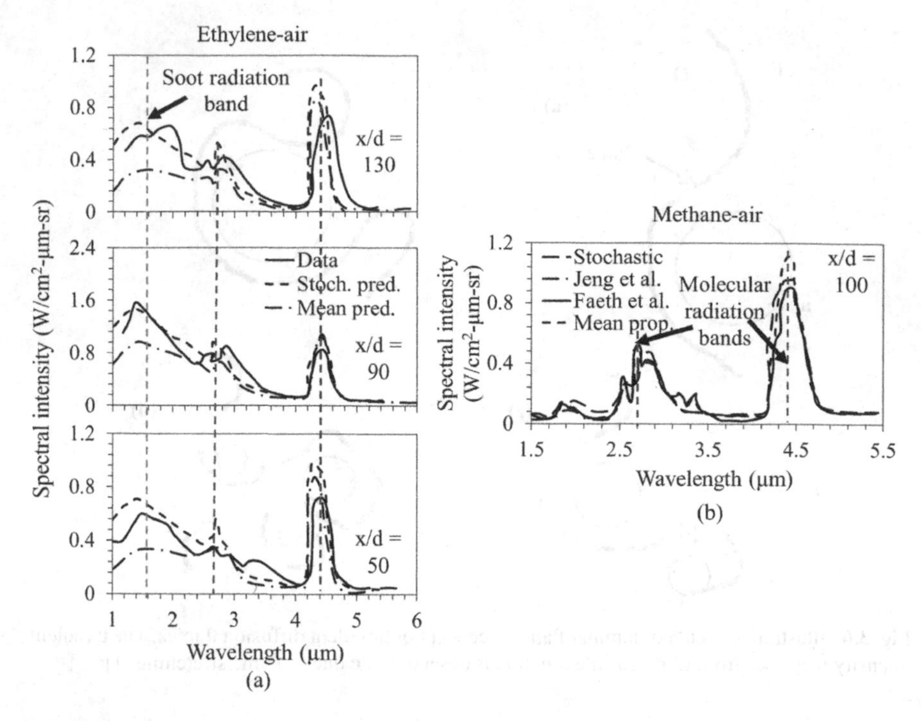

Fig. 3.7 Spectral measurements of thermal radiation from a turbulent **a** ethylene-air flame (Re = 12740) [28] and **b** methane-air flame (Re = 11700) [29]

radiation such as CO_2 and H_2O, besides the type of fuel and the residence time available. A common approach used in fire problems is identifying a radiant fraction which represents the ratio of radiant heat transfer rate from the flame to the surroundings to the total heat released by the flame. The radiant fraction is denoted as χ_R, and is expressed as

$$\chi_R = \frac{\dot{Q}_{rad}}{\dot{m}_F \Delta H_c}, \tag{3.39}$$

where $\dot{Q}_{rad}$ denotes the radiant energy transferred from the flame in the form of radiation. Based on our current understanding and knowledge-base, χ_R is estimated using empirical correlations. χ_R depends on both flame size and heat release rate and the general trend is for χ_R to increase with flame size. Depending on the fuel type and flow conditions, the radiant fraction can range from a few percent to more than 50%.

3.7 Flame Extinction

As discussed in Chap. 1, general strategies for extinguishing diffusion flames are cooling, reactant removal, chemical inhibition, and flame removal. These strategies are achieved for example, by adding sufficiently large quantities of a material (such as water) that cools the condensed fuel thereby slowing combustion, removing an essential reactant from the system, adding a chemical suppressant (such as Halon) that inhibits the reactions taking place in the flame zone, or physically removing the flame from the reactant mixture by inducing high gas velocities (for example, by explosives in large production-well fires).

The rate of gas-phase chemical reaction that was denoted by $\dot{\omega}_i'''$ in Eqs. (3.4–3.6) was thus far ignored. However, it plays an important role in flame extinction (as well as ignition). It is therefore necessary to formulate an expression to quantify $\dot{\omega}_i'''$ in order to establish critical extinction criteria. Fundamentally, $\dot{\omega}_i'''$ denoting the mass of a substance reacting in a unit volume in unit time depends on the probability of successful collisions between fuel and oxygen which increases non-linearly with temperature. A successful collision means a collision between a fuel and oxygen molecule which is violent enough to loosen and then fracture bonds which maintain the molecular structure. This proportionality was first suggested by Arrhenius in 1889 [29] by a term $e^{-E/RT}$, where E is called as the activation energy and has a value between 40 and 300 kJ/mol for most combustion reactions, R is the universal gas constant equal to 8.314 J/mol-K and T is the temperature. Figure 3.8 shows the implications of this range of E values, whereby the Arrhenius equation behaves like a mathematical switch, especially at large activation energies.

Figure 3.8 shows that as $T \rightarrow \infty, e^{-E/RT} \rightarrow 1$, meaning at this stage all collisions will lead to a reaction and the energy release is maximum. For gas-phase combustion T lies between 1500 and 2500 K as shown by the shaded area in Fig. 3.8. At 2000 K, and an activation energy of 160 kJ/mol, typical for hydrocarbon reactions, gives $e^{-E/RT} \sim 10^{-5}$. This is the approximate value of the Arrhenius term in the reaction zone for most flames. At ambient temperatures, for example, at $T = 20\,°\text{C}, e^{-E/RT} \sim 10^{-28}$! This value is very low and shows that at low temperatures the fraction of collisions that lead to successful collisions is extremely low.

A common way of expressing the reaction rate is by using an Arrhenius expression,

$$\dot{\omega}_i''' = -\rho A_0 Y_F^n Y_O^m e^{-E/RT}, \tag{3.40}$$

where ρ is the density of the gas mixture, and A_0 is called the pre-exponential constant and is around 10^{10} to 10^{15} g/m^{3}s. The fuel and oxidizer mass fractions are Y_O and Y_F, respectively, and $m + n$ is the overall order of the reaction. The negative sign denotes that species i is being lost or consumed. Table 3.4 adapted from Westbrook and Dryer [30] shows the reaction rate constants that can be used in Eq. (3.40). Since the values in Table 3.4 are in molar units, they should be converted to g/m^{3}s using Eq. (3.41).

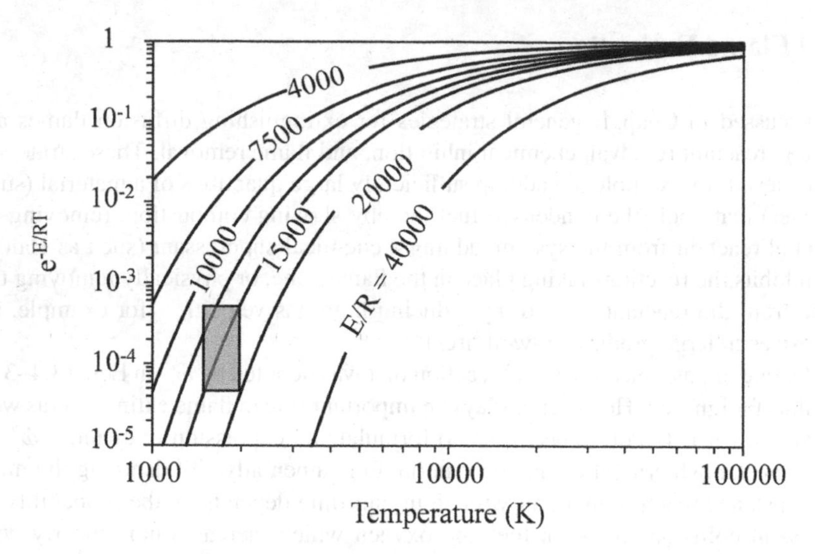

Fig. 3.8 Variation of $e^{-E/RT}$ with temperature. The shaded area shows the temperature range (1600–3000 K) found in flames. The value of the activation energy, E varies from 40 to 300 kJ/gmol for most hydrocarbon reactions resulting in E/R between 4000 K and 40000 K

Table 3.4 Overall reaction rates for fuels burning in air [30]

Fuel	Formula	B (gmol/cm^3)$^{1-m-n}$/s	E (kJ/gmol)	n	m
Methane	CH_4	8.3×10^5	125.52	−0.3	1.3
Ethane	C_2H_6	1.1×10^{12}	125.52	0.1	1.65
Propane	C_3H_8	8.6×10^{11}	125.52	0.1	1.65
n-Hexane	C_6H_{14}	5.7×10^{11}	125.52	0.25	1.5
n-Heptane	C_7H_{16}	5.1×10^{11}	125.52	0.25	1.5
Octane	C_8H_{18}	4.6×10^{11}	125.52	0.25	1.5
Methanol	CH_3OH	3.2×10^{12}	125.52	0.25	1.5
Ethanol	C_2H_5OH	1.5×10^{12}	125.52	0.15	1.6

$$\dot{\omega}_i''' = -B\rho^{m+n} \frac{Y_F^n Y_O^m e^{-E/RT}}{M_F^n M_O^m} M_F, \tag{3.41}$$

where B is in molar units (Table 3.4), and M_F and M_O represent the molecular weight of fuel and oxidizer.

The simplest possible solution is when it is assumed that order is zero, thus, the reaction term can be expressed in its simplest form as

$$\dot{\omega}_i''' = -Ae^{-E/RT}. \tag{3.42}$$

Table 3.5 Reactant consumption rate and energy consumption rate for a zeroth-order reaction with $A = 10^{13}$ g/m^{3}s and $E = 160$ kJ/mol

T (K)	E/RT	$e^{-E/RT}$	$\dot{\omega}'''_F$ (g/m^3 s)	Energy release (kJ/m^3 s)
298	64.5794	8.98551×10^{-29}	8.98551×10^{-16}	4.04348×10^{-17}
600	32.0744	1.1756×10^{-14}	0.11756	0.0052902
1200	16.0372	1.08425×10^{-7}	1.08425×10^{6}	48791.3
1600	12.0279	5.97513×10^{-6}	5.97513×10^{7}	2.68881×10^{6}
2000	9.62232	0.0000662335	6.62335×10^{8}	2.98051×10^{7}
2500	7.69786	0.000453798	4.53798×10^{9}	2.04209×10^{8}

In Eq. (3.42), A is called a pre-exponential constant. Using $A = 10^{13}$ g/m^{3}s and $E = 160$ kJ/mol, Quintiere [31] shows how the mass consumption rate of the reactant ($\dot{\omega}'''_F$) changes with temperature. The values are shown in Table 3.5 for both the consumption rate and the corresponding energy released in kW. The energy release is calculated using a typical heat of combustion of 45 kJ/g.

Table 3.5 shows that at 1200 K, the energy released in 1 cm^3 is $\sim$ 50,000 kJ/s (50 MW). This is a significant amount of energy that is released in a small volume. When temperatures are low, the energy released is really low. For example, at 600 K it is only 0.00529 kJ/s ($\sim$5 W). However, as we approach $\sim$1200 K it becomes high. The functionality of the Arrhenius expression as a mathematical switch is easily observed here. As the reaction proceeds, we use up both fuel and oxidizer. So the rate will slow down according to Arrhenius equation (Eq. (3.42)). However, in the example shown in Table 3.5 we are ignoring this by assuming that the reaction order is zero.

In Eqs. (3.40) and (3.41), the sum $(m + n)$ is called the order of the reaction. As shown in Table 3.4, the general order of reaction for most fuels, is between 1 and 2. Using these values and Eq. (3.41), and conservation equations of the species, it is also possible to include the influence of the reactants. However, the process is involved as it requires numerical integration of the conservation equations with the non-linear source term to determine the value of Y_F and Y_O at a given temperature. At 1200 K, the reaction rate is 50,000 kW/cm^3. However, this reaction is taking place because oxygen and fuel have to diffuse into the 1 cm^3 volume at a certain rate as well. Hence, the energy release per unit volume per second will be limited and ultimately a balance is achieved based on reaction and diffusion. This is why the inner region of the flame zone as shown in Fig. 3.1 is labeled the reactive-diffusive zone.

3.8 Non-Dimensionalized Energy Equation and the Damköhler Number

Once we understand, a representation of the reaction rate term, the analysis can be generalized using a non-dimensional Damköhler number [32]. To better understand the concept, let us perform an exercise of non-dimensionalizing the energy conservation equation (gas phase) using non-dimensional variables: $\hat{u} = u/u_c$, $\hat{v} = v/u_c$, $\hat{x} = x/L_c$, $\hat{y} = y/L_c$, $\hat{T} = T/T_c$, where u_c is a characteristic velocity, which we will denote as u_∞ and $L_c = L$ is some characteristic length scale. The definition of u_∞ and L, depends on the nature of the problem. For example, if the flow field is natural convective (large open pool fires, wall fire etc.) $u_\infty \sim \sqrt{gL}$ where L is the diameter of the pool fire or the length of the wall in the case of a wall fire. T_c is a characteristic temperature represented as $\Delta H_c/C_p$. Substituting the non-dimensional variables in the energy equation gives

$$\rho C_p u_\infty \left(\hat{u}\frac{\partial \hat{T}}{\partial \hat{x}} + \hat{v}\frac{\partial \hat{T}}{\partial \hat{y}} \right) = \Delta H_c \rho A Y_O^m Y_f^n e^{-E/RT} + \lambda \frac{\Delta H_c}{L^2 C_p} \left(\frac{\partial^2 \hat{T}}{\partial \hat{x}^2} + \frac{\partial^2 \hat{T}}{\partial \hat{y}^2} \right). \tag{3.43}$$

The term ($\dot{\omega}_F''' \Delta H_c$), representing the energy released due to the chemical reaction between fuel and oxidizer is expressed using an empirical Arrhenius expression (Eq. (3.40)), or specifically, the rate at which fuel is lost per unit volume. To simplify the mathematical analysis and gain physical insight into the problem, we will once again assume a zero-order reaction, simplifying the expression to $Ke^{-E/RT}$, where K is a constant with units of kg/m^3-s. Note that so far in the diffusion flame theory, we have been able to neglect this term by assuming the infinitely fast reaction assumption and the resulting Shvab-Zeldovich transformation. Rearranging terms gives

$$\left(\hat{u}\frac{\partial \hat{T}}{\partial \hat{x}} + \hat{v}\frac{\partial \hat{T}}{\partial \hat{y}} \right) = \Delta H_c (Ke^{-E/RT}) \times \frac{LC_p}{\rho C_p u_\infty \Delta H_c} + \lambda \frac{\Delta H_c}{L^2 C_p} \times \frac{LC_p}{\rho C_p u_\infty \Delta H_c} \left(\frac{\partial^2 \hat{T}}{\partial \hat{x}^2} + \frac{\partial^2 \hat{T}}{\partial \hat{y}^2} \right), \tag{3.44}$$

or

$$\left(\hat{u}\frac{\partial \hat{T}}{\partial \hat{x}} + \hat{v}\frac{\partial \hat{T}}{\partial \hat{y}} \right) = e^{-E/RT} \left[\frac{L/u_\infty}{\rho/K} \right] + \left[\frac{L/u_\infty}{L^2/\alpha} \right] \left(\frac{\partial^2 \hat{T}}{\partial \hat{x}^2} + \frac{\partial^2 \hat{T}}{\partial \hat{y}^2} \right). \tag{3.45}$$

The rearrangement of the terms as shown in Eq. (3.45) provides some interesting non-dimensional numbers. Firstly, we can define some time scales such that:

$$\tau_{\text{residence}} = \frac{L}{u_\infty} = \text{Time the reactants are present in the CV;}$$

$$\tau_{\text{chemical}} = \frac{\rho}{K} = \text{Time taken for a chemical reaction to occur;}$$

$$\tau_{\text{conduction}} = \frac{L^2}{\alpha} = \text{Pre-heat time.}$$

We can now express, the terms on the right-hand side of Eq. (3.45), in square brackets as two non-dimensional numbers which are denoted as:

$$\mathrm{Da}_1 = \frac{\tau_{\text{residence}}}{\tau_{\text{chemical}}} = \frac{L/u_\infty}{\rho/K}. \tag{3.46}$$

$$\mathrm{Da}_2 = \frac{\tau_{\text{residence}}}{\tau_{\text{conduction}}} = \frac{L/u_\infty}{L^2/\alpha}. \tag{3.47}$$

Da stands for Damköhler number, after Damköhler [32]. As Da_1 (or Da_2) decrease, suppression is obtained. This is explained using the residence and chemical times defined. $\tau_{\text{residence}}$ is the residence time which refers to the length of time the fuel vapors remain in the reaction zone. The residence time will depend on the fluid dynamics of the flame. A diffusion flame can be extinguished by the mechanism of blowout, familiar with the small flames of matches and candles. The mechanism involves the distortion of the reaction zone, within the flame in such a way as to reduce the time that the fuel vapors have to react. Blowout can occur if sufficient airflow can be achieved to reduce $\tau_{\text{residence}}$ and consequently T_f, thus reducing the Da_1 number below a critical value. This behavior is also observed in turbulent flames where with sufficiently high intensity, the laminar flamelets are distorted to an extent such that the fuel vapor does not have sufficient time to react with the oxygen. This results in further distortion of the corrugated flame illustratively shown in the transition from Fig. 3.6c–d. τ_{chemical} is the chemical reaction time, i.e., the time taken for the chemical reaction of combustion to occur. Chemical suppressants such as Halon are effective in reducing the rate of gas-phase reaction. This increases the chemical time, thereby decreasing Da_1 and leading to extinction.

If the conduction losses are increased (denominator of Da_2) extinction can be achieved by quenching due to heat losses (in-depth conduction, re-radiation, and radiative feedback from the flame). For example, if the fuel supply can be decreased by say lowering the vaporization rate of a condensed fuel by adding water, the flame approaches the surface and thus loses more heat which may ultimately result in quenching due to increase in heat losses. The quenching due to lack of fuel supply has not been studied extensively with condensed fuels and works by Roberts and Quince [33] and Torero et al. [34] represent the few studies that define a minimum fuel supply stability limit for diffusion flames established over a condensed fuel. Torero et al. [34] show that for diffusion flames spreading on condensed fuel surfaces, such as upward spread along walls, or horizontal spread or lateral spread on floors, the mechanism of gas-phase flame quenching is dependent on blowout mechanisms (reduction of Da_1) for the leading edge (location of flame anchored to the surface) and quenching (reduction of Da_2) for the trailing edge (location of the flame tip).

Extinction of diffusion flames can also be physically explained using a critical flame temperature concept. The S-curve first studied by Fendell [35] and Liñan [36] is usually used to represent the response of temperature to the system Damköhler number. For most fire problems the critical flame temperature at which a diffusion flame ceases to exist is approximately 1300 °C [37, 38]. However, this value should be used with caution. A recent study by T'ien and Endo [39] shows that there is no unique critical extinction temperature for different materials and the flame temperature at extinction is a strong function of flow parameters.

3.9 Premixed Flames

Fire mainly involves laminar and turbulent diffusion flames. So far we discussed the combustion process in fire problems for example flames like a candle, jet, pool, and solid fuels. In such flames the reaction velocity or reaction rate of the process that produces heat is dependent on the mixing process which is laminar or turbulent diffusion. In a premixed flame on the other hand, the flame is a surface where fresh unburned gas and the burned mixture meet and which pushes forward to the fresh unburned gas. The flame thus has a speed or velocity associated with it which is related to the rate at which fresh unburned gas can be converted to the burned mixture. The study of premixed flame thus involves analyzing the mechanisms responsible for the conversion of the fresh gas to the burned gas at the flame. Since the velocity of the flame is a direct consequence of the mechanism, the study of premixed flame relates the burning velocity to these mechanisms. Typical values of hydrocarbon air flames range from 1 to 100 cm/s with flame thickness of the order of 1 mm. The highest velocity is that of H_2—air mixtures which can be upto 300 cm/s (6.3 mph), or around an average jogging speed.

Premixed flames are important precursors and their study is necessary to prevent a potential fire hazard. In order to have a diffusion flame, a premixed flame must first exist, for example, at the source of initial ignition where a flame kernel has to form and propagate to some length. In regions where a diffusion flame is adjacent to a cold wall or where the diffusion flame is anchored to the fuel surface, there exists premixed flame behavior. Pockets of premixed flames also exist in turbulent diffusion flames. Such flames are called partially premixed flames where both diffusion and premixing play an important role in flame dynamics.

Figure 3.9 shows a schematic of the primary difference between diffusion and premixed flame where the fuel and oxidizer denoted by *F* and O, respectively, arrive at the flame front on opposite sides for diffusion and on the same side for premixed flames. The difference is clearly observed, for example, with say a spherical fuel mass as shown in Fig. 3.9 in an oxidizing environment. Because the fuel and oxygen are not mixed, the mass of fuel will shrink with time as combustion proceeds. On the other hand, if the fuel and oxygen were premixed and ignited, the flame propagation would proceed outward in search of fresh fuel-oxidizer mixture. The propagation rate in the two cases will also be much different. Highlighting this difference and deriving an expression for the fuel-oxidizer mixture consumption rate as a function of concentration, temperature, and pressure forms the basis of premixed flame theory.

Figure 3.10a shows a combustible mixture contained in a tube and ignited by a spark. A premixed flame propagates through the mixture moving along the tube at a velocity characteristic to the mixture.

The laminar burning velocity is a fundamental thermokinetic property of the mixture giving an indication of the rate of fuel consumption at the reaction zone or flame surface. The laminar burning velocity is defined as the velocity of the unburned gas approaching normal to the flame surface. In the Bunsen burner it can be easily measured by the sine of the half angle of flame cone multiplied by the exit velocity

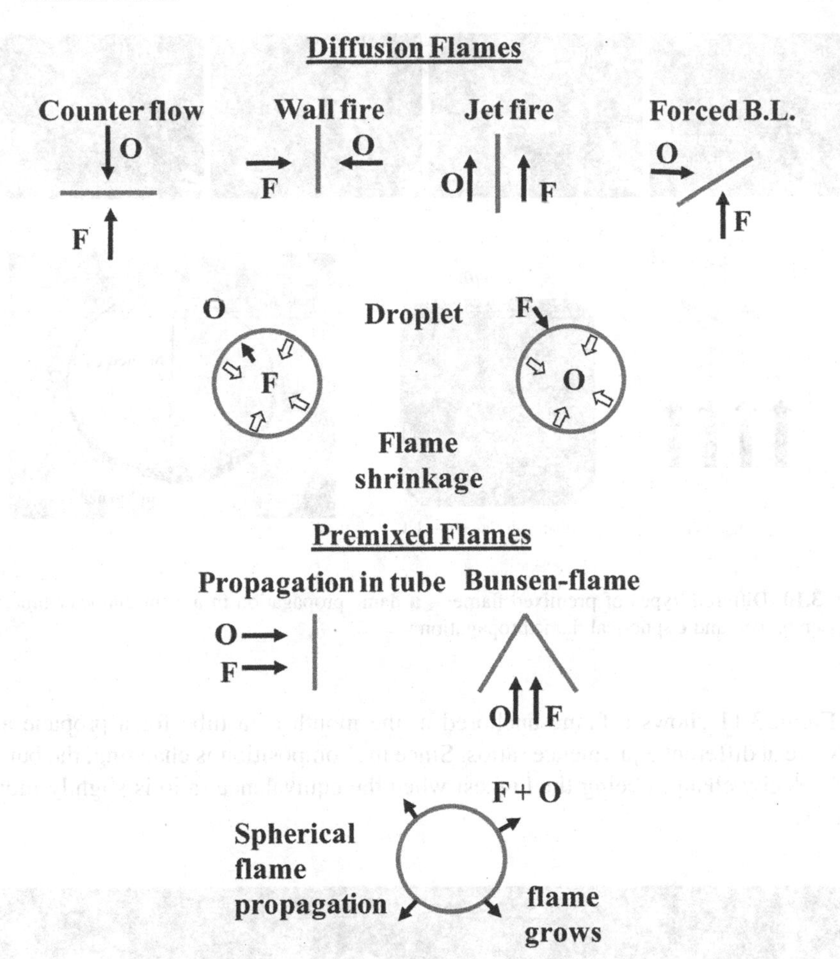

Fig. 3.9 Schematic showing difference between diffusion and premixed flames

of the gas leaving the tube as shown in Fig. 3.10b. Hence the burning velocity does not change with the increase or decrease of the flow speed. Increasing the flow velocity in the tube causes the cone to elongate and the angle to decrease. If the flow velocity is decreased below the burning velocity, the flame flattens and then propagates down the tube. This is called flashback. Similarly, at high flow velocity the flame can no longer elongate to accommodate the high flow rate and loses its anchor at the base. This is called blow off. Figure 3.10c shows a combustible gas-mixture ignited in an explosion sphere with the spark at the center. The spherical flame propagates outwards in search of fresh fuel leaving the burned mixture within as shown in Fig. 3.10c. In all three cases the velocity depends on the pressure, initial temperature, and its chemical composition. The effect of composition on the laminar burning velocity is shown in Fig. 3.11.

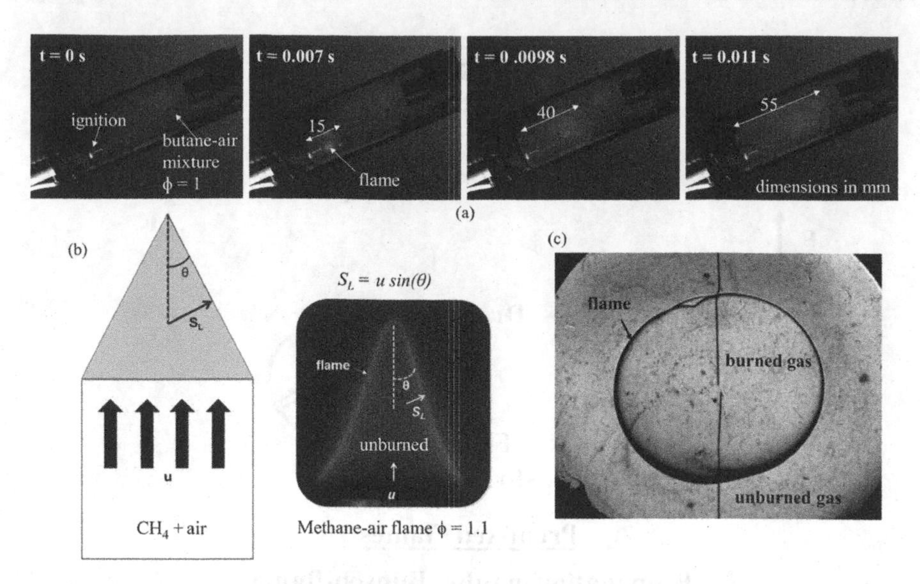

Fig. 3.10 Different types of premixed flames—**a** flame propagation in a 5 cm diameter tube, **b** Bunsen burner, and **c** spherical flame propagation

Figure 3.11 shows a flame anchored at the mouth of a tube for a propane air mixture at different equivalence ratios. Since the composition is changing, the burning velocity changes being the highest when the equivalence ratio is slightly more than 1.

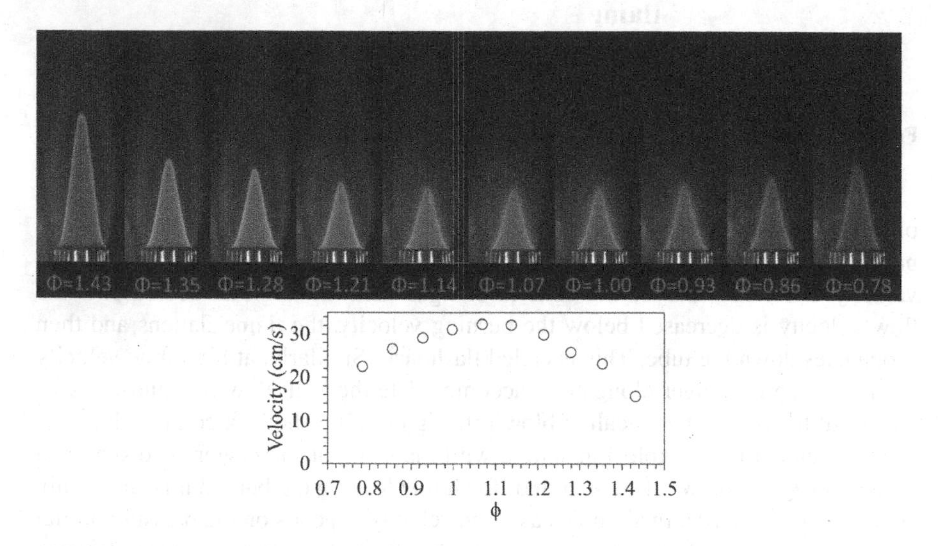

Fig. 3.11 Methane-air premixed flame at different equivalence ratios and laminar burning velocity using the cone angle method

3.10 Structure of a Premixed Flame

The picture of the premixed flame is obtained by solving for the spatial distribution of pressure, velocity, temperature, and concentration of all species (Fig. 3.12). Such comprehensive theories are presented in reference books on combustion. In Fire, it is important to identify the hazard of a fuel oxidant mixture, and this does not need the detailed flame structure. For most practical problems, for example, design of a flame arrestor (quenching of a flame), flammability limits, and minimum ignition energy, experimental data from a standard experiment, and a combination of empirical and theoretical expressions are usually used.

The propagation of gas mixtures in tubes led to the early theories on premixed flames, by Coward and Hartwell [40, 41] who found that a horizontal tube (5–10 cm diameter) when filled with a methane-air mixture with 10% methane and ignited at the open end, formed a flame that moved along the tube with a velocity which they described as equal to

$$S_L = \frac{\text{Volume flow rate}}{\text{Area of flame front}},$$

and found that was a constant irrespective of the propagation direction: horizontal, upward, or downward orientation of the tube.

Assuming steady flame propagation, conservation of mass across the flame front gives $\rho_u S_L = \rho_b u_b$, where subscript u denotes the unburned gases and subscript b denotes the burned gas, giving $u_b = \frac{\rho_u S_L}{\rho_b} = \frac{T_b}{T_u} S_L$.

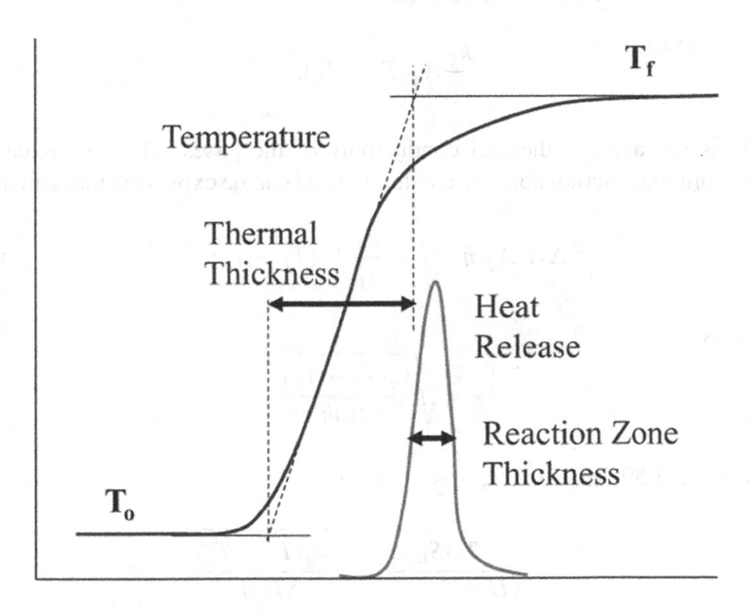

Fig. 3.12 Structure of a premixed flame

For most hydrocarbons $S_L \sim 0.3$ to 0.8 m/s except for H_2—air mixtures where it has a maximum of 3 m/s and C_2H_2-air having a maximum of 1.5 m/s. Also, for most hydrocarbons, $\frac{\rho_u}{\rho_b} \sim \frac{T_b}{T_u} \sim \frac{T_f}{T_u} \sim 10$. Thus, the velocity of the burned gases leaving the flame is ~5 m/s. The value of S_L depends on the equivalence ratio and peaks at close to but not exactly 1. The tabulated values of S_L in standard textbooks normally refer to a composition at which it is maximum. Nearly 50 years before the experiments by Coward and Hartwell [40], Mallard and Le Chatelier (1883) [42] developed a thermal theory which was the first theoretical expression to the laminar burning velocity.

Theory of Mallard and Le Chatelier The heat release of a premixed flame can be expressed as:

$$\dot{Q}_f = \Delta H_c A_f \dot{m}''' \delta_f. \tag{3.48}$$

In Eq. (3.48), ΔH_c is the heat of reaction (kJ/kg), A is the area of the flame (m^2), $\dot{m}'''$ is the mass consumption rate (kg/m^{3}s) and δ_f is the flame thickness (mm). If the heat released is used completely to raise the temperature of gases from T_u to T_f then,

$$\Delta H_c A_f \dot{m}''' \delta_f = C_p(T_f - T_u) S_L \rho_u A_f, \tag{3.49}$$

and the flame thickness can then be equated as,

$$\delta_f = \frac{C_p(T_f - T_u) S_L \rho_u}{\Delta H_c \dot{m}'''}. \tag{3.50}$$

The heat conducted by the flame zone is

$$\frac{k_g}{\delta_f} A_f (T_f - T_u),$$

where k_g is the average thermal conductivity of the gases. The heat released is removed from the reaction zone by conduction and can be expressed mathematically as,

$$\Delta H_c A_f \dot{m}''' \delta_f = \frac{k_g}{\delta_f} A_f (T_f - T_u), \tag{3.51}$$

which gives

$$\delta_f = \sqrt{\frac{k_g (T_f - T_u)}{\Delta H_c \dot{m}'''}}. \tag{3.52}$$

Equating Eqs. (3.50) and (3.52) gives

$$\frac{C_p(T_f - T_u) S_L \rho_u}{\Delta H_c \dot{m}'''} = \sqrt{\frac{k_g (T_f - T_u)}{\Delta H_c \dot{m}'''}}, \tag{3.53}$$

giving

$$S_L = \frac{\Delta H_c \dot{m}'''}{C_p(T_f - T_u)\rho_u}\sqrt{\frac{k_g(T_f - T_u)}{\Delta H_c \dot{m}'''}}. \tag{3.54}$$

For a purely adiabatic system where there are no losses, the heat of combustion can be written as $\Delta H_c = C_p(T_f - T_u)$ and thus Eq. (3.54) can be simplified to

$$S_L = \sqrt{\frac{k_g \dot{m}'''}{\rho_u^2 C_p}} = \sqrt{\frac{\alpha \dot{m}'''}{\rho_u}}, \tag{3.55}$$

where $\alpha = k/(\rho C_p)$ is the thermal diffusivity. Equation (3.55) is useful for our purposes as it shows the controlling parameters affecting the laminar burning velocity. It is seen from Eq. (3.55) that S_L increases with the square root of the reaction rate and the thermal diffusivity. ρ is proportional to P, and $\dot{m}'''$ is proportional to P^n, where n is the order of the reaction. For a bimolecular collision n equals 2, and for most hydrocarbon reactions n takes an empirical value of 1.8.

This makes S_L proportional to $P^{\frac{n}{2}-1} \sim P^{-0.1}$ and the burning velocity decreases slightly as the pressure rises. Also, δ_f is proportional to $P^{-\frac{n}{2}}$ and hence low pressure flames are thick. Equation (3.55) also shows that if the specific heat is high, the laminar burning velocity is lowered. This makes sense because the specific heat is the amount of energy needed to raise the temperature by 1 °C. So higher the specific heat (C_p) of the gas mixture more is the energy needed to raise the temperature of the gas. Since the chemical reaction rate follows the Arrhenius expression discussed earlier ($e^{-E/RT}$) which behaves as a mathematical switch triggering at a critical temperature, more energy has to be fed into the system with a higher specific heat. Thus, S_L reduces with increasing C_p. Ultimately, the thermal theory can be summarized simply by formulating three balances:

1. A convective—diffusive balance given by

$$C_p(T_f - T_u)S_L\rho_u A_f = \frac{k_g}{\delta_f}A_f(T_f - T_u)$$

2. A reactive-diffusive balance given by

$$\Delta H_c A_f \dot{m}'''\delta_f = \frac{k_g}{\delta_f}A_f(T_f - T_u)$$

3. A global energy balance given by

$$\Delta H_c = C_p(T_f - T_u)$$

The three balances yield Eq. (3.55). It is interesting to note the primary difference between the simplified theory of Mallard and Le Chatelier and the more comprehensive theories discussed in reference textbooks. The main approximation lies in the fact that δ_f comprises both the pre-heat or thermal and chemical reaction zone thickness shown in Fig. 3.12. We assume that the reaction is occurring everywhere! This is why Eq. (3.55) actually over predicts experimental data. However, all the dependencies are correct. The over-prediction was resolved by Zeldovich, Frank-Kamenetskii, and Semenov by introducing a Zeldovich number in the expression. Another interesting aspect with premixed flames is the interplay between thermochemistry and chemical kinetics. The expression for the laminar burning velocity,

$$S_L = \sqrt{\frac{\alpha \dot{m}'''}{\rho_u}},$$

is dependent mainly on the rate of consumption of reactants in the flame zone $\dot{m}'''$. However, the adiabatic flame temperature can be simply obtained by solving $\Delta H_c = C_p(T_f - T_u)$. Thus, the controlling parameters for adiabatic flame temperature are not the same as that of S_L.

3.11 Quenching of a Premixed Flame

The concept of a quenching distance or the minimum distance between two surfaces such that a flame is not able to pass through has been used since the invention of the Davy safety lamp in coal mines, where the screen size around a flame is kept at a quenching distance so that when used in an explosive coal mine, ignition is avoided.

Quenching is related to losses, and quenching the flame can be visualized by considering failure of a flame to propagate in a pipe of diameter d. If the thickness of the flame is t_f as shown in Fig. 3.13, the heat generated in the flame (kW) is given by

$$\dot{q}_{gen} = \frac{\pi d^2}{4} t_f \dot{q}''', \tag{3.56}$$

where $\dot{q}'''$ is the heat released per unit volume due to the combustion reaction. The rate of heat loss from the flame to the wall is:

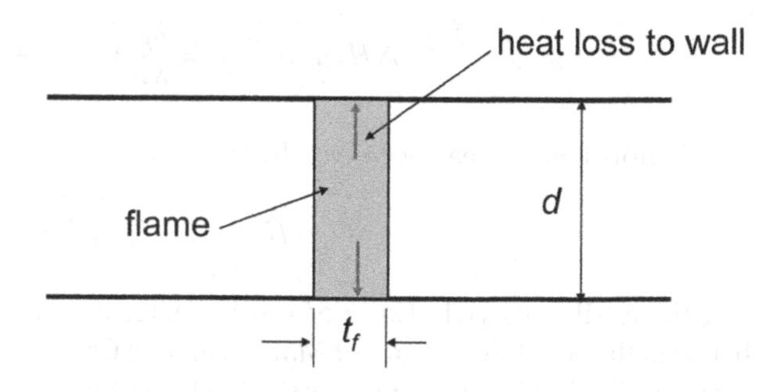

Fig. 3.13 Quenching of flame from heat loss to the walls

$$\dot{q}_{loss} = k_g(\pi d t_f)\frac{dT}{dr}, \tag{3.57}$$

where k_g is the mean thermal conductivity of the combustion products in the width t_f, T is the temperature and r is the radial coordinate.

The temperature gradient $\left(\frac{dT}{dr}\right)$ can be approximated as:

$$\frac{dT}{dr} = \frac{T_f - T_w}{d/2}, \tag{3.58}$$

where T_f is the flame temperature and T_w is the temperature of the wall which is assumed to be equal to ambient. Quenching of the flame will take place when the generation rate is less than the heat loss rate. The threshold diameter for quenching (quenching distance d_q) can be found by equating the heat loss (Eq. (3.57)) with heat generated (Eq. (3.56)) to give:

$$\frac{\pi}{4} d_q^2 t_f \dot{q}''' = k_g \pi d_q t_f \frac{T_f - T_w}{d_q/2}. \tag{3.59}$$

In Eq. (3.59), d_q is the quenching diameter or quenching distance. Equation (3.59) can be simplified further to give,

$$\frac{1}{4} d_q \dot{q}''' = k_g \frac{T_f - T_w}{d_q/2},$$

$$d_q = \sqrt{\frac{8k_g(T_f - T_w)}{\dot{q}'''}}. \tag{3.60}$$

The primary constraint is the heat generation rate per unit volume ($\dot{q}'''$). Rearranging Eq. (3.54) and noting that $\Delta H_c \dot{m}''' = \dot{q}'''$ and $T_u \approx T_w$,

$$\dot{q}''' = \frac{\rho^2 C_p^2 S_L^2 (T_f - T_w)}{k_g} = \frac{S_L^2 (T_f - T_w) k_g}{\alpha^2}. \tag{3.61}$$

And, the quenching diameter is equal to

$$d_q = \sqrt{\frac{8k_g(T_f - T_w)}{k_g(T_f - T_w)S_L^2/(\alpha^2)}} = \frac{2\sqrt{2}\alpha}{S_L}. \tag{3.62}$$

The quenching of a premixed flame is thus inversely related to the laminar burning velocity. A highly energetic gas-air mixture, for example, H_2-air will correspondingly have a low quenching distance and hence is also more hazardous. The quenching distance is used in the design of flame arrestors installed to prevent flame propagation into vent pipes, ducts, and openings of storage vessels. The device consists of a multitude of narrow channels or gaps. Typical quenching distances of ~ 1 to 2 mm are sufficient for most hydrocarbon air flames.

The maximum gap in the arrester mesh to prevent flame propagation for most hydrocarbon fuels is approximately half of the quenching distance. For example, the quenching distance for gasoline vapors is approximately 2 mm, making the maximum safe gap for an arrester screen for storage vessels of gasoline approximately 1 mm. These values, reported as the maximum experimental safe gap (MESG) are tabulated for multiple fuels in codes and standards like NFPA 497.

3.12 Flammability Limits (LFL and UFL) and Limiting Oxygen Concentration (LOC)

Figure 3.11 shows that the laminar burning velocity varies with equivalence ratio, whereby it is maximum for an equivalence ratio slightly greater than 1, but for lean (~50% stoichiometric) and rich mixtures (~3 times stoichiometric) its value drops. In fact less than or more than these critical values, it is not possible to sustain a self-propagating flame. Such values of ϕ_{min} or ϕ_{max} establish the flammability limits of a fuel-air mixture. Since S_L is a function of composition, initial temperature, and pressure; a flammability limit can be defined as the limit of one of these quantities beyond which a fuel-oxidizer mixture cannot be made to burn. The flammability limits are usually tabulated in codes and standards. Extensive tabulation can be found in Coward and Jones [43] and Zabetakis [44] based on experiments performed by the Bureau of Mines in 1950–1965.

An interesting point to note is that LFL is independent of O_2/N_2 ratio in air. For example, if all the N_2 is replaced by O_2, the LFL will not change. This is because both O_2 and N_2 have the same specific heat and thus the heat of combustion transferred to either gas will lead to the same reaction temperature. The UFL on the other hand depends strongly on the concentration of O_2 in air, since this limit is caused by an excess of fuel, i.e., by a deficiency of oxygen. Bartknecht [45] has also shown that the relative humidity has a perceptible influence on the width of the flammability range, with the widest range observed in extremely dry mixtures.

Flammability limits are of practical interest in safety considerations because mixtures outside of these limits can be handled without concern for ignition. LFL and UFL are based on fuel in air and a theoretical limit can be derived by making use of the fuel concentration in the expression for reactivity of fuel-air mixture in W/m^3-s using the Arrhenius one-step global reaction model described earlier. Such theoretical expressions are available in Annamalai and Puri [46]. For most practical applications, though, the flammability limits can be correlated using the heat of combustion as shown by Suzuki and Koide [47, 48]:

$$LFL = \frac{-3.42}{\Delta H_c} + 0.569\Delta H_c + 0.0538\Delta H_c^2 + 1.8,$$
$$UFL = 6.30\Delta H_c + 0.567\Delta H_c^2 + 23.5. \quad (3.63)$$

In Eq. (3.63), 123 organic materials containing carbon, hydrogen, oxygen, nitrogen, and sulfur were analyzed. The heat of combustion is in kJ/mol and the LFL and UFL are in vol% of fuel in air.

Adding an increasing quantity of inert gas or a non-combustible mixture can also lead to a critical concentration causing a no flame propagation. This is usually expressed in terms of a limiting oxygen concentration (LOC) and is based on the concentration of inert that needs to be added to prevent ignition of a combustible gas-air mixture. The efficiency of the inert gas is based on the quantity of heat transferred during the flame propagation and ensuring that this heat is absorbed without causing further reaction to sustain flame propagation. The order of efficiency of the inert gas added is the same as the heat capacity.

For instance, to estimate the Limiting Oxygen Index (LOC) for butane (C_4H_{10}): The lower flammability limit (LFL) for butane is = 1.9% by volume. The overall chemical reaction of butane at stoichiometric condition is

$$C_4H_{10} + 6.5O_2 \rightarrow 4CO_2 + 5H_2O.$$

The LOC can be estimated as

$$\text{LOC} = \left[\frac{\text{moles fuel}}{\text{total moles}}\right]_{LFL} \left[\frac{\text{moles } O_2}{\text{moles fuel}}\right]_{stoic} = 1.9\frac{6.5}{1} = 12.4 \text{ vol \% } O_2.$$

3.13 Minimum Ignition Energy (MIE)

The minimum ignition energy is the minimum energy input required to initiate combustion. All flammable vapor (including dusts) have a MIE. Many hydrocarbons have MIE of 0.25 mJ. This is low enough to cause ignition by electrostatic discharges. Walking across a rug on a cold winter day initiates a static discharge of around 22 mJ. An ordinary spark plug has a discharge of 25 mJ. Thus, given a flammable gas-mixture ignition is easy. MIE decreases with increase in pressure. Pressure is the net force that gas molecules are exerting when they collide with the walls of a container. So a high pressure means more number of collisions per second of gas molecules on the container and/or each collision is occurring with a greater force. Thus gas molecules in a high-pressure condition are moving about very fast, thus have a higher energy. This means they will need lesser energy from an ignition source to react. The MIE is related to the burning velocity by

$$MIE = \frac{k_g}{S_L}(T_f - T_o).$$

The autoignition temperature is different from the minimum ignition energy and denotes the minimum temperature at which a gas mixture will react spontaneously without the presence of a spark. It depends on the concentration or equivalence ratio of the gas mixture, pressure, flow condition, initial temperature and the presence/absence of a catalyst. The ignition of a vapor is also possible by adiabatic compression. For example, vapors in a gasoline engine ignite when compressed to a temperature that exceeds AIT creating an engine knock. Several large accidents have been caused due to flammable vapors sucked into the intake of air compressors. The example below adapted from Crowl and Louvar [49] illustrates this mode of ignition. For instance, a lubricating oil has an AIT of 400 °C, the compression ratio required to raise the temperature of air to the AIT of this oil can be evaluated.

Assuming an initial air temperature of 25 °C,

$$T = P^{\frac{\gamma-1}{\gamma}} \Rightarrow P = T^{\frac{\gamma}{\gamma-1}}.$$

where, γ is the ratio of specific heat C_p/C_v and is equal to 1.4 for air

$$\Rightarrow \frac{P_f}{P_i} = \left(\frac{T_f}{T_i}\right)^{\frac{\gamma}{\gamma-1}} = \left(\frac{400+273}{25+273}\right)^{\frac{1.4}{0.4}} = 17.3$$

Therefore compression ratio should be kept below 17.3 × 14.7 psia = 254 psia. Lubricating oil in piston type compressors is always found in minute amounts in the cylinder bore. Compressor operations must always be kept below the AIT of the oil to prevent explosion. The various parameters that have been discussed in this chapter are tabulated in Table 3.6 for different fuels.

3.14 Summary of Diffusion and Premixed Flames

A flame is a surface in the gas phase where a rapid change occurs in a thin layer accompanied by generation of heat. We discussed two types of flames that can exist in practice in this chapter. One is a diffusion flame or a non-premixed flame, for example, a candle flame where fuel vapor rises from the wick and burns in the neighborhood of the wick only to the extent that it is mixed with oxygen from the air. The heat generated is used not only to make the oxygen and fuel vapor reach a temperature where they can react chemically but also to provide heat for the candle wax to melt and vaporize. In the candle flame the rate of generation of heat is thus determined solely by the transport of fuel vapor and mixing of the vapor with oxygen from air. In larger fires, for example, say a sofa on fire, the increase in scale causes turbulence which causes the mixing process to increase more than the laminar diffusion flame of a candle and produces more heat. This is best illustrated using a jet flame where a greater velocity tends to lengthen the flame. However, once turbulent mixing occurs, the mixing process increases and causes the flame length to shorten. This reduction

Table 3.6 Flame and flammability characteristics for different fuels

Fuel	T_{ad}	S_L	MIE	d_q	AIT	LFL	UFL	ϕ_{LFL}	ϕ_{UFL}	LOC	LOC	r_{st}	ΔH_C	ΔH_C
-	K	cm/s	mJ	mm	K	% Fuel	% Fuel	-	-	N_2-air	CO_2-air	% Fuel	kJ/mol	kJ/g
	[50]	[51]	[52]	[52]	[52]	[44]	[44]	[53]	[53]	[54]	[54]	[52]	[52]	[52]
H_2	2400	312	0.018	0.55	673	4	75	0.14	2.54	4.6	4.6	29.5	241.8	119.96
CO	2370	46	–	1.73	882	12.5	74	0.34	6.76	5.1	5.1	29.5	283	10.1
CH_4	2226	40	0.28	2.5	810	5	15	0.46	1.64	11.1	13.1	9.47	802.3	50.1
C_2H_2	2541	166	0.017	0.55	578	2.5	100	0.19	∞	NA	NA	7.74	1255.5	48.22
C_2H_4	2370	80	0.09	1.25	763	2.7	36	0.41	6.1	8.5	10.2	6.53	1323.1	47.16
C_2H_6	2260	42.5 [55]	0.25	2	745	3	12.4	0.5	2.72	9.5	11.9	5.65	1428.6	47.5
C_3H_8	2257	46	0.26	2.10	743	2.1	9.5	0.51	2.5	10.7	12.8	4.02	2043.1	46.3
C_4H_{10}	2260	45	0.26	2.20	638	1.8	8.4	0.57	2.84	10.6	13.0	3.12	2656	45.7

T_{ad}—Adiabatic flame temperature; S_L—Laminar burning velocity; MIE—Minimum ignition energy

d_q—Quenching distance; AIT—Autoignition temperature; LFL—Lower flammability limit

UFL—Upper flammability limit; ϕ - Equivalence ratio; LOC - Limiting oxygen concentration

r_{st}—Stoichiometric volume concentration of fuel; ΔH_C—Heat of combustion

in flame height of the turbulent non-premixed jet flame is the most conclusive proof that turbulent mixing and not diffusion will determine the energy release rate for larger scale fires.

The second class of flames is those with premixed gases. Here we deal with a completely different situation where the fuel and oxygen are perfectly mixed at the very onset. While in a diffusion flame, the flame is the surface where the fuel and oxygen meet stoichiometrically and react, in a premixed flame the flame is the surface where fresh gas and burned mixture meet and which pushes forward into the fresh gas. The flame thus has a speed or velocity associated with it which forms a fundamental thermokinetic property of the mixture called laminar burning velocity. The study of premixed flames thus boils down to study of mechanisms responsible for the conversion of fresh gas to burned gas at the flame. Since the velocity of the flame is a direct consequence of the mechanism which can be experimentally measured in an easy way, the theory of premixed flames relates the burning velocity to these mechanisms.

3.15 Case Study

Consider a geometrical configuration identical to that of the one-dimensional problem presented earlier, with a mixture of 22% oxygen and 78% nitrogen as oxidizer and a mixture of 80% Ethane (C_2H_6) and 20% nitrogen as fuel (all percentages are in volume). The mixture fraction, the fuel and oxygen concentrations, and the temperature as a function of x and the flame location (x_f) can be determined for a given value of U.

Part 1: Mixture Fraction Assume thermal properties as those of air at 1000 K, a plot of $Z = \frac{(e^{x/\delta}-1)}{(e^{1/\delta}-1)}$ is shown in Fig. 3.14. The value for δ based on U of 1 mm/s equals 0.168 m. It is assumed that the thermal diffusivity is that of air at 1000 K ($\alpha = 168 \times 10^{-6}$ m^2/s).

Part 2: Flame Location The flame is located at the position where $\beta_{OF} = \frac{Y_{O,i}}{s} - Y_F = 0$. This corresponds to a stoichiometric mixture. The flame location can be found using Eq. (3.23). $Y_{O,i} = \frac{0.22\times 32}{0.22\times 32+0.78\times 28} = 0.244$ and $Y_{F,i} = \frac{0.8\times 30}{0.8\times 30+0.2\times 28} = 0.811$, where, the molecular weight of ethane is 30 g/mol. $s = \frac{Y_O}{Y_F}_{stoic}$ is obtained assuming a one step overall reaction of ethane reacting with oxygen and equals $s = 112/30 = 3.73$. Substituting these values in Eq. (3.23) gives

$$x_f = 0.168 \ln \left(\frac{\frac{0.244}{3.73}(e^{\frac{1}{0.168}} - 1)}{0.811 + \frac{0.244}{3.73}} + 1 \right) = 0.57.$$

The location of the flame is shown in Fig. 3.15. It occurs at a location where the concentration of fuel (Y_F) and oxidizer (Y_O) is zero. Note that since the chemical

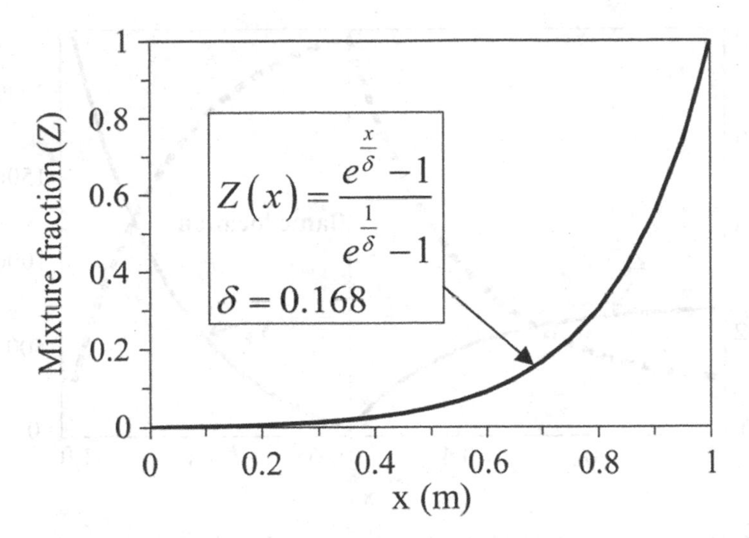

Fig. 3.14 Mixture fraction (Z) as a function of x

reaction is assumed to be infinitely fast, there is no fuel or oxidizer leakage on either side of the flame. Further, the flame location occurs at the location of maximum temperature. Profiles of fuel, oxidizer, and temperature are evaluated in Part 3 and 4 discussed next.

Part 3 and 4: Profiles of Oxygen, Fuel, and Temperature First assume that there is no oxygen in the fuel zone and no fuel in the oxidizer zone. Use β_{OF} to find the fuel and oxygen mass fractions.

$$x < x_f : \quad Y_O = s\beta_{OF}(x) \text{ and } Y_F = 0;$$

$$x > x_f : \quad Y_O = 0 \text{ and } Y_F = -\beta_{OF}(x).$$

Since the oxygen and fuel mass fractions are known, the temperature can be plotted using β_{FT} or β_{OT}.

$$\beta_{FT} = Z(x)[\beta_{FT}(1) - \beta_{FT}(0)] + \beta_{FT}(0)$$
$$= \left(\frac{e^{x/\delta}-1}{e^{1/\delta}-1}\right)\left[Y_{F,i} + \frac{T_i}{(\Delta H_c/C_p)} - \frac{T_i}{(\Delta H_c/C_p)}\right] + \frac{T_i}{(\Delta H_c/C_p)}.$$

Substituting β_{FT} from Table 3.2, we can solve for $T(x)$ as:

$$T(x) = (\Delta H_c/C_p)$$
$$\times \left[\left(\frac{e^{x/\delta}-1}{e^{1/\delta}-1}\right)\left[Y_{f,i} + \frac{T_i}{(\Delta H_c/C_p)} - \frac{T_i}{(\Delta H_c/C_p)}\right] + \frac{T_i}{(\Delta H_c/C_p)} - Y_F(x)\right].$$

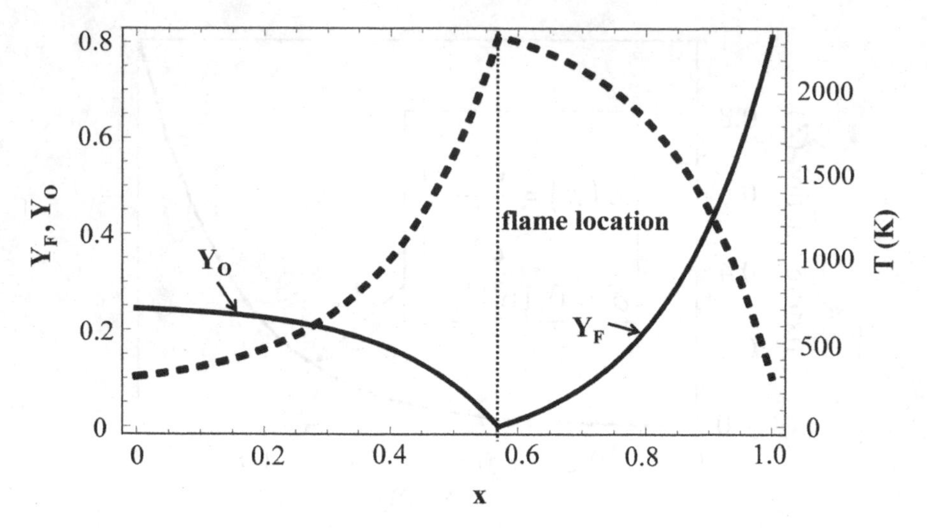

Fig. 3.15 Mass fraction profiles of fuel and oxygen and temperature profile

The specific heat, $C_p = 1.14$ J/g at 1000 K and the heat of combustion is equal to $\Delta H_c = 47500$ J/g. The temperature profile is shown in Fig. 3.15.

The characteristic length scale δ represents the ratio of the mass diffusion to the velocity of the gas stream. This characteristic distance represents the extent that heat and mass is diffusing against the gas flow into the gas stream. The example shows how the combustion in a diffusion flame is determined by the diffusional and hydrodynamic supply of the oxidizer and fuel. It uses constant surface boundary conditions at fuel and oxidizer side. For the case of condensed fuels at the interface between fuel and air, specifying a constant mass fraction for the species as the boundary condition at the interface provides great simplicity in formulating an analytical solution. However, determining an approximate value of $Y_{F,x=\text{surface}}$, the fuel mass fraction at the interface, is complicated as it involves solving a heat and mass balance at the fuel surface. Further complication is that typically the species concentrations are discontinuous at the interface between two materials, whereas temperature is continuous.

To take an example, consider a heptane pool fire. If we are interested in determining the rate at which heptane vapor is transferred to the gas phase, we would need to specify the vapor concentration of heptane in the gas-phase side of the heptane-air interface. The mass fraction of heptane inside the pool is unity (neglecting the small amount of oxygen or nitrogen dissolved in heptane). However, it would be incorrect to assume $Y_{F,x=\text{surface}} = 1$ for the gas-phase mass fraction of heptane vapor at the interface. This value primarily depends on the interface temperature besides the pressure. Interface temperature is determined by the heat balance analysis between gas phase to interface. The conditions at the interface are based on relationships that are theory based or deduced from experiments.

An additional non-dimensional number called as B-number arises during the solution of establishing the mass fraction of fuel vapor at the interface. The B-number is an important derived property, which is dependent on the thermo-physical properties of the system, that provides an expression for the mass burning rate eventually. Further details are discussed in Chaps. 4 and 5.

Review Questions

1. Describe the major difference between a premixed and a non-premixed flame and categorize the following into premixed and diffusion flame: (a) flame over a fuel droplet, (b) Bunsen burner flame, and (c) forest fire.
2. Determine Prandtl number, Schmidt number and Lewis number for CO_2 at pressure (P) and temperature (T) of $P = 1$ atm, $T = 298$ K, when it diffuses in air. Also compare the mass diffusivity when: (a) $P = 5$ atm, $T = 298$ K, and (b) $P = 1$ atm, $T = 1800$ K.
3. Consider an ethanol pool with its surface maintained at a known temperature ($295\ K <= T <= 345$ K) using a radiant heater. The fuel mass fraction decays to zero at a distance (L) away from the surface of the fuel pool given as a function of surface temperature as $L(m) = 0.0013T - 0.26$. Determine the variation of fuel mass fraction with surface temperature (T) at a distance of $L = 12$ cm from the pool surface. Also determine the minimum fuel surface temperature such that a flame is possible at this height (LFL = 3.3% by volume of air).
4. What does the value of smoke point of a fuel signify?
5. How does the assumption of constant velocity and infinitely fast reaction simplify the problem of determining the diffusion flame location? Also specify the physical significance of unity Lewis number?
6. What is the difference between coupling function and mixture fraction?
7. Why is the estimation of thermo-physical properties difficult while solving fire problems?
8. Using simple scaling analysis for obtaining time for convection and diffusion, the flame height for a jet diffusion flame is observed to be a function of jet diameter (d), velocity (V) and the diffusion coefficient (D). Plot the variation of flame height with the following parameters:

   (a) Varying volume flow rate of fuel.
   (b) Varying fuel inlet temperature.
   (c) Varying burner diameter at constant fuel inlet velocity.

9. How are turbulent eddies responsible for formation of flamelets?
10. Plot the variation of the fuel consumption rate assuming zeroth-order reaction with:

    (a) varying temperature from 300 K to 1500 K, for activation energy of 100 kJ/mol.

(b) varying activation energy from 100–300 kJ/mol, at temperature of 1500 K. Assume pre-exponential factor to be 1013 g/m^{3}s.

11. Explain in brief, using definitions of Damköhler number, if any of the following techniques could lead to fire suppression: (a) increasing the characteristic velocity, (b) heating the ambient air, (c) cooling blanket around the fuel pool, (d) radiation dome to re-radiate heat from the flame to the pool.
12. Differentiate between flashback and blow off.
13. Consider a Bunsen burner with diameter d with fuel-air mixture entering at a constant volume flow rate. The flame height (h_d) is measured for each diameter and it is observed to vary as $h_f = 7.14(d) - 2.8$, where h_f and d are in cm. Determine and plot the variation of laminar burning velocity with the burner diameter.
14. Using the Theory of Mallard and Le Chatelier, evaluate the laminar burning velocity of hydrogen at equivalence ratio of 1, when the consumption rate is 2 mol/cm^3-s. Evaluate air properties at an average temperature, assuming unburnt gas temperature to be 300 K.
15. Determine the quenching diameter for methane-air flame at unburnt temperature of 298 K and pressure of 2 atm. The laminar burning velocity (S_L) of methane-air mixture is seen to be vary as a function of pressure P (atm) as S_L (cm/s) $= 43P^{-0.5}$.
16. For a given hydrocarbon C_xH_y, determine an expression for limiting oxygen concentration in terms of x, y and its heat of combustion.
17. How does the minimum ignition energy vary with the equivalence ratio of the fuel-air mixture?

References

1. Rangwala, A. S. 2015. Diffusion flames, *SFPE handbook of fire protection engineering*, 350–372, Springer.
2. Arias-Zugasti, M., and D.E. Rosner. 2008. Soret transport, unequal diffusivity, and dilution effects on laminar diffusion flame temperatures and positions. *Combustion and Flame* 153 (1): 33–44.
3. Smyth, K.C., J.H. Miller, R.C. Dorfman, W.G. Mallard, and R.J. Santoro. 1985. Soot inception in a methane/air diffusion flame as characterized by detailed species profiles. *Combustion and flame* 62 (2): 157–181.
4. Spalding, D.B. 1950. Combustion of liquid fuel in gas stream. *Fuel* 29: 2–7.
5. Spalding, D. B. 1955. *Some fundamentals of combustion*, Gas Turbine Series, vol. 2. London: Academic, Butterworths Scientific Publications.
6. Chakrabarty, R.K., H. Moosmüller, W.P. Arnott, M.A. Garro, G. Tian, J.G. Slowik, E.S. Cross, J.H. Han, P. Davidovits, T.B. Onasch, and D.R. Worsnop. 2009. Low fractal dimension cluster-dilute soot aggregates from a premixed flame. *Physical Review Letters* 102 (23): 235–504.
7. Kent, J., and H.G. Wagner. 1984. Why do diffusion flames emit smoke. *Combustion Science and Technology* 41 (5–6): 245–269.
8. De Ris, J., and X.F. Cheng. 1994. The role of smoke point in material flammability testing. *Proceedings of Fire Safety Science* 4: 301–312.

9. Markstein, G. H. 1989. Correlations for smoke points and radiant emission of laminar hydrocarbon diffusion flames. In *Proceedings of combustion institute*, vol. 22(1) 363–370. Elsevier.
10. Delichatsios, M. 1993. Smoke yields from turbulent buoyant jet flames. *Fire Safety Journal* 20 (4): 299–311.
11. Lautenberger, C.W. 2002. CFD simulation of soot formation and flame radiation (MS dissertation, Worcester Polytechnic Institute).
12. Zeldovich, I., G.I. Barenblatt, V. Librovich, and G. Makhviladze. 1985. *Mathematical theory of combustion and explosions*. Moscow: Nauka.
13. Shvab, V.A. 1948. *Goz*. Moscow-Leningrad: Energ. izd.
14. Zel'dovich, Y. B. 1950. Zhur. Tekhn. Fiz, English translation: NACA Tech, Memo No. 1296, 1949. 19: p. 1199.
15. Burke, S.P., and T.E.W. Schumann. 1928. Diffusion flames. *Industrial Engineering Chemistry* 20 (10): 998–1004.
16. Rangwala, A.S., V. Raghavan, J.E. Sipe, and T. Okano. 2009. A new property evaluation scheme for mass transfer analysis in fire problems. *Fire Safety Journal* 44 (4): 652–658.
17. Law, C.K., and F.A. Williams. 1972. Kinetics and convection in the combustion of alkane droplets. *Combustion and Flame* 19: 393–405.
18. Roper, F. 1977. The prediction of laminar jet diffusion flame sizes: Part I Theoretical model. *Combustion and Flame* 29: 219–226.
19. Roper, F. 1978. Laminar diffusion flame sizes for curved slot burners giving fan-shaped flames. *Combustion and Flame* 31: 251–258.
20. Turns, S.R. 2000. *An introduction to combustion*. New York: McGraw-Hill.
21. Sunderland, P., J. Quintiere, G. Tabaka, D. Lian, and C.-W. Chiu. 2011. Analysis and measurement of candle flame shapes. *Proceedings of the Combustion Institute* 33 (2): 2489–2496.
22. Bilger, R. 1989. Turbulent diffusion flames. *Annual Review of Fluid Mechanics* 21 (1): 101–135.
23. Peters, N. 2000. *Turbulent combustion*. Cambridge University Press.
24. Libby, P. A., and F. A. Williams 1994. *Turbulent reacting flows*. Academic.
25. Peters, N. 1984. Laminar diffusion flamelet models in non-premixed turbulent combustion. *Progress in Energy and Combustion Science* 10 (3): 319–339.
26. Williams, F. 1975. Recent advances in theoretical descriptions of turbulent diffusion flames, *Turbulent mixing in nonreactive and reactive flows*, 202–208.
27. Kuznetsov, V. 1982. *The effect of turbulence on the formation of large superequilibrium concentrations of atoms and free radicals in diffusion flames*, 3–9. Akademiia Nauk SSSR: Izvestiia, Mekhanika Zhidkosti i Gaza.
28. Gore, J. and G. Faeth 1988. Structure and spectral radiation properties of turbulent ethylene/air diffusion flames, Proceedings of Combustion Institute 21(1): 1521–1531. Elsevier.
29. Arrhenius, S. 1889. On the reaction rate of the inversion of non-refined sugar upon souring. *Z Phys Chem* 4: 226–248.
30. Westbrook, C.H., and F.L. Dryer. 1981. Simplified reaction mechanisms for the oxidation of hydrocarbon fuels in flames. *Combustion Science and Technology* 27: 31–43.
31. Quintiere, J.G. 2006. *Fundamentals of fire phenomena*. New York: Wiley.
32. Damköhler, G. 1947. The effect of turbulence on the flame velocity in gas mixtures.
33. Roberts, A., and B. Quince. 1973. A limiting condition for the burning of flammable liquids. *Combustion and Flame* 20 (2): 245–251.
34. Torero, J.L., T. Vietoris, G. Legros, and P. Joulain. 2002. Estimation of a total mass transfer number from the standoff distance of a spreading flame. *Combustion Science and Technology* 174 (11): 187–203.
35. Fendell, F.E. 1965. Ignition and extinction in combustion of initially unmixed reactants. *Journal of Fluid Mechanics* 21 (2): 281–304.
36. Liñan, A. 1974. The asymptotic structure of counterflow diffusion flames for large activation energies. *Acta Astronautica* 1 (7): 1007–1039.
37. Rasbash, D.J. 1962. The extinction of fires by water sprays. *Fire Research Abstracts and Reviews* 4 (1): 28–53.

38. Quintiere, J., and A. Rangwala. 2004. A theory for flame extinction based on flame temperature. *Fire and Materials* 28 (5): 387–402.
39. T'ien, J. S. and M. Endo. 2013. Material flammability: A combustion science perspective. Proceedia Engineering 62:120–129.
40. Coward, H., and F. Hartwell. 1932. Studies in the mechanism of flame movement Part I: The uniform movement of flame in mixtures of methane and air, in relation to tube diameter. *Journal of the Chemical Society* 1996–2004.
41. Coward, H., and F. Hartwell. 1932. Studies in the mechanism of flame movement Part II: The fundamental speed of flame in mixtures of methane and air. *Journal of the Chemical Society* 2676–2684.
42. Mallard, E. 1883. Recherches experimentales et theoriques sur la combustion des melanges gaseux explosifs. Annals mines 8(4):274–568.
43. Coward, H. F. and G. W. Jones. 1952. Limits of flammability of gases and vapors, vol. 503. US Government Printing Office.
44. Zabetakis, M. G. 1965. Flammability characteristics of combustible gases and vapors. Bureau of Mines Washington DC.
45. Bartknecht, W. 1989. *Explosions: course, prevention, protection*. Berlin, New York: Springer.
46. Annamalai, K. and I. K. Puri. 2006. *Combustion science and engineering*. CRC Press.
47. Suzuki, T., and K. Koide. 1994. Correlation between upper flammability limits and thermochemical properties of organic compounds. *Fire and Materials* 18 (6): 393–397.
48. Suzuki, T. 1994. Quantitative structure-property relationships for auto-ignition temperatures of organic compounds. *Fire and Materials* 18 (2): 81–88.
49. Crowl, D. A. and J. F. Louvar. 2001. *Chemical process safety: fundamentals with applications*. Pearson Education.
50. Morley, C. 2005. *Gaseq: A Chemical Equilibrium Program for Windows - Ver. 0.79*.
51. NFPA 68. 2013. *Standard on explosion protection by deflagration venting*. National Fire Protection Association.
52. Babrauskas, V. 2003. *Ignition handbook*, - Vol. 8027. WA, Issaquah: Fire Science Publishers.
53. Barnett, H. C. and R. R. Hibbard. 1958. Basic considerations in the combustion of hydrocarbon fuels with air - Vol. 1300. US Government Printing Office.
54. NFPA 69. 2007. *Standard on explosion prevention systems*. National Fire Protection Association.
55. Mitu, M., D. Razus, V. Giurcan, and D. Oancea. 2014. Experimental and numerical study of laminar burning velocity of ethane-air mixtures of variable initial composition, temperature and pressure. *Energy and fuels* 28 (3): 2179–2188.

Chapter 4
Burning of Liquids

Several industries and laboratories handle many types of flammable liquids and use them as solvents and chemicals. The occurrences of small flame over an accidentally spilled fuel surface and its growth to a larger fire have been encountered by many industries. The petroleum industry that handles crude oil has encountered several scenarios of accidental crude oil spills. The best way to clean up this oil spill is to burn it in an in-situ manner. Liquid fuels have also been used to create pool fires for research purposes in labs and certain industries.

The calorific value of a liquid fuel governs the amount of heat released during its burning. Other than this, a few more properties are used to characterize a liquid fuel. The vaporization of a liquid fuel is dictated by its volatility, which is governed by properties such as boiling point, latent heat of vaporization, and specific heat. The ignition of the liquid fuel, and formation of a flame over its surface are dictated by its flash and fire points. Flash point of a liquid fuel is the minimum temperature of the liquid at which sufficient vapors are produced, which mix with atmospheric air and produce a flash or an instantaneous flame, when a pilot flame is introduced. That is, the flash point corresponds to the formation of a momentary premixed flame. Liquid fuels can be divided into various classes based on their flash point (FP). Spillage of Class I fuels (FP $< 38\,°C$) is extremely hazardous with regard to spread of fire, spillage of Class II fuels ($38\,°C < FP < 60\,°C$) is considered dangerous from fire point of view, only if an ignition source is present. Class III fuels (FP $> 60\,°C$) are generally safe, unless a very large heat source, capable of raising the fuel temperature significantly, exists. Fire point is a temperature higher than the flash point. At fire

A. S. Rangwala and V. Raghavan, *Mechanism of Fires*,
https://doi.org/10.1007/978-3-030-75498-3_4

point, when a pilot ignition source is introduced, a flame is established over the liquid surface, which sustains even after the removal of the pilot ignition source. Boiling point is a temperature higher than the fire point and it is the saturation temperature of the liquid at the given pressure. The flash and fire point temperatures are not unique values as the boiling point. They lie in a certain range based on the type of apparatus (open cup apparatus or closed cup apparatus) used to measure their values and the ambient conditions.

The latent heat of vaporization is the energy that is required for converting the liquid to its vapor at the saturation temperature for a given pressure. However, sensible heating is required to raise the temperature of the fuel to the saturation temperature before this phase change can occur. The sensible heat is evaluated using the specific heat of the fuel, which is defined as the energy required per unit mass of fuel to raise the temperature by unit degree Celsius or Kelvin.

Rate of flame propagation on any liquid fuel surface, and its ultimate growth to a larger fire, primarily depends on the above-mentioned properties of the fuel, the ambient conditions, and the heat feedback from the flame by conduction, convection, and radiation. The process of flame spread over a liquid fuel surface is transient (unsteady) in nature. When the temperature of the liquid fuel exceeds its fire point, a diffusion flame is formed over its surface upon ignition. This flame starts spreading over the fuel surface. The rate of flame spread depends on the gas-phase convective flow. Based on the direction of the convective flow relative to the direction of the spread, the flame spread is classified as concurrent or opposed-flow flame spread. The liquid-phase transport processes are generally important in the analysis of flame spread along with the gas phase. After spreading over the fuel surface completely, when the flame is sustained over the entire area of the fuel surface, it consumes the liquid fuel at an almost steady rate depending on the thickness of the liquid fuel layer. This is called mass burning rate. Under steady burning regime, the liquid fuel surface reaches an equilibrium temperature, which is close to its boiling point. The gas-phase transport processes primarily control the mass burning rate. The mass burning rate further depends on the prevailing atmospheric conditions, forced and natural convective air flow. Convective flow relative to a steadily burning fuel surface can be classified into cross-flow, opposed-flow, and co-flow configurations. In a practical fire scenario, many of the above configurations in the steady burning regime are encountered. The transient flame spread process and the subsequent steady burning characteristics under several conditions are of interest and important due to safety implications.

4.1 Evaporation

Evaporation of the liquid fuel is the first step in its burning process. In liquid fuel pool fire scenarios, generally the heat from the ambient (flame) is incident over the fuel surface (interface between the liquid and gas phase). This heat is partly transferred into the liquid phase (based on the thickness of the pool) and partly used to evaporate

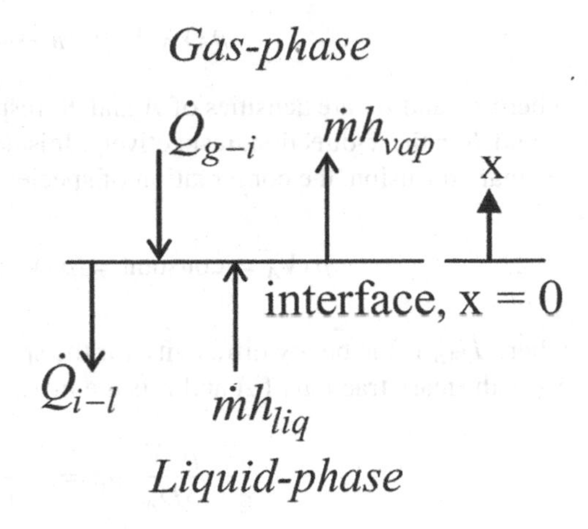

Fig. 4.1 Heat and mass transport processes at the liquid interface

the liquid fuel. The mass evaporated $\dot{m}$ is transferred into the gas phase. This coupled mass and energy transport processes are shown in Fig. 4.1, where the subscripts *g-i* represents gas phase to interface and *i-l* represents interface to liquid phase.

The energy balance is written as

$$\dot{Q}_{g-i} - \dot{Q}_{i-l} = \dot{m}(h_{vap} - h_{liq}) = \dot{m}h_{fg}, \tag{4.1}$$

where $\dot{Q}_{g-i}$ is the heat transferred from gas to the interface, $\dot{Q}_{i-l}$ is the heat transferred from interface to the liquid, h_{vap} is the enthalpy of saturated vapor, h_{liq} is the enthalpy of saturated liquid, and h_{fg} is the latent heat of vaporization of the liquid fuel, which is a weak function of temperature and usually taken as a constant.

To analyze steady evaporation of a liquid, consider a liquid fuel, A, contained in a vessel in an ambient, where a mixture of gas B and vapors of liquid A flow in the direction parallel to the liquid surface. This is shown in Fig. 4.2. Mass fraction of vapor, A, ($Y_{A,\infty}$) flowing across is less than that at the interface ($Y_{A,i}$). A steady evaporation is attained by supplying the liquid at the rate at which it evaporates. This assures that the fuel does not regress and the interface remains at the same x-location. The overall mass conservation for this problem in the x-direction, is given as

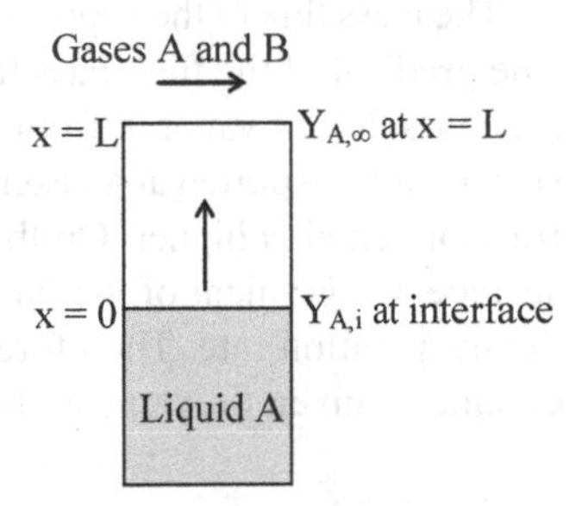

Fig. 4.2 Schematic of steady evaporation of a liquid into a binary mixture

$$\rho_A V_A + \rho_B V_B = \text{constant}, \tag{4.2}$$

where ρ_A and ρ_B are densities of A and B, respectively, V_A and V_B are velocities of A and B in the x-direction, respectively. It is noted that V_B is zero. By Fick's law of ordinary diffusion, the conservation of species A is written as

$$\rho_A V_A = \text{constant} = \rho_A V_A Y_A - \rho D_{AB} \frac{dY_A}{dx}, \tag{4.3}$$

where D_{AB} is the binary diffusivity for the species pair A and B, expressed in m^2/s, Y_A is the mass fraction of A and ρ is the mixture density. Rearranging the variables,

$$\frac{\rho_A V_A}{\rho D_{AB}} dx = -\frac{dY_A}{1 - Y_A}. \tag{4.4}$$

Assuming the product, ρD_{AB} is constant with respect to x, the above expression can be integrated to obtain

$$\frac{\rho_A V_A}{\rho D_{AB}} x = \ln(1 - Y_A) + \ln C. \tag{4.5}$$

Using the boundary condition, $Y_A(x = 0) = Y_{A,i}$, the constant, C, can be eliminated to obtain

$$Y_A(x) = 1 - (1 - Y_{A,i}) \exp\left(\frac{\rho_A V_A}{\rho D_{AB}} x\right). \tag{4.6}$$

It is clear that the mass flux of species A depends on the interface mass fraction of vapor A, $Y_{A,i}$, which further depends on the surface temperature. The mass flux can be calculated by using the far boundary condition, that is, at $x = L$, $Y_A = Y_{A,\infty}$, which is given as

$$\rho_A V_A = \frac{\rho D_{AB}}{L} \ln\left(\frac{1 - Y_{A,\infty}}{1 - Y_{A,i}}\right) = \frac{\rho D_{AB}}{L} \ln(1 + B_Y). \tag{4.7}$$

In Eq. (4.7), $(1 + B_Y) = (1 - Y_{A,\infty})/(1 - Y_{A,i})$. The parameter, B_Y, is called Spalding's transfer number.

The mass flux of the vapor from the interface is driven by the gas-phase conditions. The gradient of the fuel mass fraction at the interface, as shown in Eq. (4.3), dictates the mass flux of vapor A from the interface. If the vapor formed over the liquid fuel surface is transported at a faster rate into the gas phase, then the gradient of fuel mass fraction remains higher. On the other hand, if the vapor accumulates over the pool surface, the gradient of the fuel mass fraction decreases and results in a decrease in the evaporation rate. Therefore, evaporation in a convective condition is seen to be enhanced, given the same ambient conditions.

The evaporation can be analyzed using energy balance that is given in Cartesian coordinates as,

$$\dot{m}'' C_p \frac{dT}{dx} = k_g \frac{d^2T}{dx^2}, \tag{4.8}$$

where $\dot{m}'' = \dot{m}''_A = \rho_A V_A$ and k_g is the thermal conductivity of the gas phase mixture adjacent to the interface. Considering steady evaporation ($\dot{m}''$ is constant) and constant properties (C_p and k_g), integration of Eq. (4.8) results in,

$$\frac{dT}{dx} = (Z\dot{m}''T + c_1). \tag{4.9}$$

Here, $Z = C_p/k_g$ and c_1 is an integration constant. Separating the variables and integrating Eq. (4.9) yields

$$\frac{1}{Z\dot{m}''} \ln(Z\dot{m}''T + c_1) = x + c_2. \tag{4.10}$$

The constants, c_1 and c_2, can be evaluated using the boundary conditions. Application of the boundary condition at $x = L$, $T = T_\infty$, in Eq. (4.10) yields

$$\frac{1}{Z\dot{m}''} \ln(Z\dot{m}''T_\infty + c_1) = L + c_2. \tag{4.11}$$

Combining Eqs. (4.10) and (4.11), after application of the other boundary condition at $x = 0$, $T = T_s$, in Eq. (4.10), yields,

$$\ln(Z\dot{m}''T_s + c_1) = -Z\dot{m}''L + \ln(Z\dot{m}''T_\infty + c_1). \tag{4.12}$$

The integration constants, c_1 and c_2, can be evaluated by using Eqs. (4.11) and (4.12). The expression for the temperature profile along the x-direction is given as

$$T(x) = \frac{(T_\infty - T_s)e^{Z\dot{m}''(x-L)} - T_\infty e^{-Z\dot{m}''L} + T_s}{1 - e^{-Z\dot{m}''L}}. \tag{4.13}$$

When steady evaporation occurs, the heat balance at the liquid fuel surface is given as

$$\dot{m}'' h_{fg} = k_g \left. \frac{dT}{dx} \right|_{x=0}. \tag{4.14}$$

From the temperature profile in Eq. (4.13), the derivative of the temperature at $x = 0$ is,

$$\left. \frac{dT}{dx} \right|_{x=0} = \frac{Z\dot{m}''(T_\infty - T_s)e^{-Z\dot{m}''L}}{1 - e^{-Z\dot{m}''L}}. \tag{4.15}$$

Substitution of Eq. (4.15) in Eq. (4.14) yields

$$\dot{m}''h_{fg} = k_g \frac{Z\dot{m}''(T_\infty - T_s)e^{-Z\dot{m}''L}}{1 - e^{-Z\dot{m}''L}}. \tag{4.16}$$

Simplification of Eq. (4.16) yields

$$\dot{m}'' = \frac{k_g}{C_p L} \ln(B_q + 1). \tag{4.17}$$

Here, Spalding's transfer number is given as

$$B_q = C_p(T_\infty - T_s)/h_{fg}. \tag{4.18}$$

The mass evaporation flux obtained by Eq. (4.17) should be same using Eq. (4.7) under steady evaporation conditions. Based on the ambient temperature, the liquid fuel surface attains an equilibrium temperature, close to the saturation temperature, at the steady state. That is, thermodynamic equilibrium prevails at the interface. Here, the partial pressure of the fuel vapor over the liquid surface is the saturation pressure of the liquid at the surface temperature. The vapor pressure is calculated using a thermodynamic relationship at a given temperature. This can be an equation such as Wagner's equation or the Clausius-Clapeyron equation or any other thermodynamic relationship (NIST database). This assumption is valid for the evaporation and combustion occurring at low to moderate ambient pressures. For an ambient temperature lesser than the boiling point of the liquid, the equilibrium temperature is slightly lesser than the ambient temperature, as a process called evaporative cooling occurs. If the ambient temperature is higher than the boiling point, the equilibrium temperature attains a value slightly lesser than the boiling point. This means that even when a flame steadily sustains over a liquid pool surface, the equilibrium temperature of the interface will be slightly lesser than the boiling point. This equilibrium temperature remains almost a constant along the interface during a steady evaporation or burning process. The equilibrium temperature as a function of ambient temperatures and corresponding evaporation rates for the three fuels—methanol, *n*-Heptane, and *m*-Xylene are discussed in case study 1 in Sect. 4.5.

4.2 Ignition

In this section, phenomenon involved during the ignition of a liquid fuel surface is explained. A heat source provides the necessary heat for the liquid surface to reach its flash point temperature [1] from its initial temperature. It should be noted that the fuels with flash point temperature lower than the ambient temperature, such as gasoline, *n*-Heptane, and methanol, will have enough vapors adjacent to their surface to mix with the ambient air, without the need of an additional heat source. The reactant mixture formed over the liquid surface is ignited by an ignition source and a momentary premixed flame is formed. After this flash formation, it takes a while to form the reactant mixture again and to cause a premixed flame to show up with the help of the ignition source. It is important for the occurrence of ignition that the reactant mixture (fuel vapor + air) formed over the liquid surface is flammable (exists within the flammable limits defined for the fuel), as well as, it has enough volume. When a heat source with sufficient energy (called minimum ignition energy) is introduced to this mixture, ignition takes place. That is, for piloted (using an external heat source) ignition to occur, the requirements are the presence of minimum volume of a flammable mixture, and an ignition source with sufficient energy. The ignition location also plays an important role. The ignition location cannot be too close to the fuel surface as that region may not have enough oxygen. Similarly, it cannot be too far. In fact, in standard test procedure carried out to determine the ignition delay, the igniter is kept at a particular location above the fuel surface.

As the fuel surface is further heated to its fire point, enough vapors are available to sustain a flame over the liquid surface. This flame is a diffusion or non-premixed flame, where the fuel vapor is transported from the liquid surface to the flame and the air is transported from the ambient to the flame zone. As a result of this, certain quantity of heat is released. If the quantity of heat released is more than the heat losses to the ambient, then the combustion becomes sustainable. The ignition source may now be removed, as the diffusion flame is capable of providing the necessary heat to the fuel surface, which in turn provides enough fuel vapors to sustain the flame. It is difficult to ignite a liquid pool with a high fire point using a pilot flame at a given location on the pool surface. The pilot flame creates gas-phase movement as well as temperature gradient over the liquid surface. This imparts convection currents due to shear stress and surface tension gradients. As a result, the heat applied to the surface quickly dissipates and delays the ignition process. A wick placed over the liquid surface facilitates the ignition process as it arrests the convective motion of the liquid surface. Generally the material used as a wick has low thermal conductivity that prevents heat losses. The fuel is heated up to its fire point sooner and a diffusion flame is formed within a shorter induction time. The other method is to use small amount of high volatility liquid with lower density over the fuel surface and ignite it. This method is used in igniting large oil spills in the ocean during the process of in-situ burning as a means of clean up.

When the fuel vapor and air mixture is exposed to a temperature called autoignition temperature, the mixture is ignited without the need of an external ignition source. A common example for this is the flame formation on the surface of over-heated cooking oil kept on a frying pot that is being heated underneath. In practice, the liquid fuel or flammable solvent coming into contact with a hot surface of a machinery component evaporates and produces vapor at a higher temperature that mixes with air. Temperature of the mixture sometimes exceeds the autoignition temperature and, depending upon the temperature of the hot surface, the mixture can self-ignite. In general, autoignition temperature of several fuels is higher than their boiling point. However, for fuel like n-dodecane the autoignition temperature (478 K) is less than its boiling point (489 K). In such cases, autoignition may not occur [2, 3]. Autoignition temperatures are determined by a standard test using ASTM E659 Standard apparatus under idealized conditions.

Self-heating and spontaneous combustion of high boiling point vegetable oils in porous substrate such as fabric materials are also encountered. Self-heating occurs due to the changes in the structure of the fuel and causes smoldering through the fuel bed. Both the porous substrate (if combustible) and the liquid undergo slow oxidation process. It is reported that the self-heating hazard is of increasing order for coconut oil, olive oil, cotton seed oil, soybean oil, sunflower oil, linseed oil, and tung oil, indicated by the increasing values of their iodine numbers. The iodine number is the mass of iodine in grams that is consumed by 100 g of oil. The idea is based on the fact that the reactivity of a substance to oxygen from atmosphere depends on the number of double bonds which can be attacked by oxidizing agents and thereby produce local heat.

The ignition time, also called the ignition delay, is the sum of evaporation time, mixing time, and the induction time. The evaporation time depends on the rate of evaporation, controlled by heat and mass transfer processes. The time required for mixing of the reactants is determined by the flow field and is usually much lesser in a convective environment, especially when the flow is turbulent. Induction time is defined as the time for the commencement of chemical reactions and is a function of the temperature. Both piloted ignition and autoignition conditions reduce the induction time nearly to zero. Thus, it is generally the evaporation time that controls the ignition delay.

4.3 Flame Spread over Liquid Fuel Surfaces

The transient flame propagation over liquid fuel surface depends on the fuel properties, initial fuel temperature, ambient conditions, and the strength and direction of the convective flow. As discussed earlier, the primary fuel properties, which dictate the flame propagation, are the flash and the fire points. The flame propagation can be classified as concurrent flame propagation or opposed flame propagation, depending upon the direction of flame spread with respect to that of the oxidizer flow. In concurrent flame spread, oxidizer flows in the same direction as that of the propagating flame and in opposed flame spread, oxidizer flow is opposite to the flame spread direction. In both cases, the oxidizer flow could be induced by forced and/or natural convection mechanisms. Concurrent flame spread is highly rapid and could be hazardous. In opposed flame spread, the heat transfer from the flame to the fuel surface and interface mass transfer processes are affected by the oxidizer flow, which is against the direction of the propagating flame. Consequently, the rate of flame propagation is generally low and the flame front is well defined.

The rate of flame propagation in both the concurrent and opposed-flow modes is also influenced by whether the fuel is above or below its flash point temperature. When the liquid fuel is at a temperature above its flash point, under equilibrium conditions, combustible mixture of fuel and oxidizer is present everywhere above the fuel surface. Ignition of this mixture causes the flame to propagate directly over the fuel surface in a manner similar to that in a premixed combustible gas. Here, the

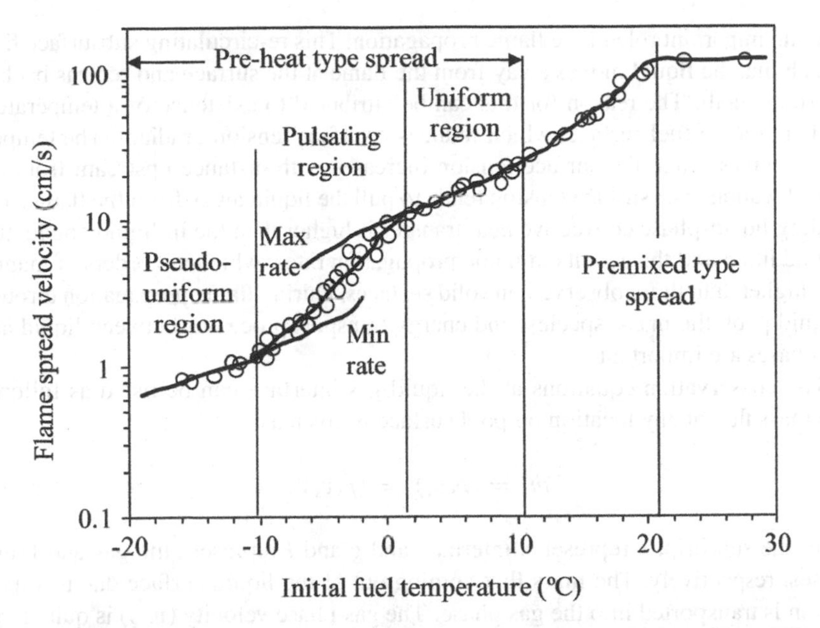

Fig. 4.3 Flame spread over methanol as a function of initial temperature (Adopted from Akita [4])

flame propagation is basically controlled predominantly by events in the gas phase and the spread rates are very high (~1 m/s). Once the flame has propagated over the entire liquid surface, the liquid pool merely acts as a source of fuel to sustain the steady burning phase.

When the liquid fuel is at a temperature below its flashpoint, the mass transfer process will be stronger and rate controlling. For a liquid fuel with its temperature lower than its flash point, sufficient amount of fuel vapor—air mixture within the flammability range will not be available above the fuel surface. Therefore, the propagating flame has to first pre-heat the downstream liquid fuel surface to its fire point temperature, such that a flammable mixture is generated for further flame propagation. In this case, liquid-phase transport processes will also become significant controlling mechanisms. The spread rates in these cases are appreciably slower. Figure 4.3 presents the variation of flame spread rate as a function of initial temperature of the methanol pool (Flash point = 15 °C, Boiling point = 64 °C), as reported in [4].

Various regimes appear as the initial temperature is increased from sub-zero temperature to ambient temperature of 30 °C. The flame spread has been observed to fall under uniform regime, pulsating regime, and pseudo-uniform regime, when the fuel temperature is below the flash point. It transitions from uniform regime, showing a significant increase in its spread rate, to a regime where flame spreads at a constant speed, encountered in the flame propagation through gaseous reactant.

Studies on flame propagation over liquid fuels at sub-flash temperatures have revealed that the fuel does not remain stationary beneath the flame. A re-circulatory convection current is observed in the liquid fuel surface ahead of the flame, which plays an important role in the flame propagation. This re-circulating subsurface flow is such that the liquid moves away from the flame at the surface and returns back at a certain depth. The reason for this can be attributed to existence of a temperature gradient on the fuel surface, which induces a surface tension gradient. The temperature decreases and the surface tension increases with distance upstream from the flame location. The surface tension tends to pull the liquid away from the flame. The ensuing liquid-phase convective heat transfer is higher than the in-liquid conductive heat addition and this results in flame propagation rates, which are orders of magnitude higher than those observed on solid surfaces. During flame propagation through a liquid pool, the mass, species, and energy transport processes between liquid and gas phases are important.

The conservation equations at the liquid-gas interface can be listed as follows. The mass flux at any location on pool surface is given as:

$$\dot{m}'' = (\rho_s v_s)_g = (\rho_s v_s)_l. \tag{4.19}$$

Here, the subscript s represents interface and g and l represent the gas and liquid phases, respectively. The mass flux coming out of the liquid surface due to vaporization is transported into the gas phase. The gas-phase velocity ($v_{s,g}$) is quite large when compared to the liquid-phase velocity ($v_{s,l}$) due to the large difference between the densities of liquid and gas phases ($\rho_{s,l}$ and $\rho_{s,g}$, respectively). The mass of the fuel vapor is transported by convection and diffusion within the corresponding phases, as dictated by Fick's law. In case of two component fuel or a polar-alcohol liquid fuel like methanol, which can absorb water, the conservation of fuel includes diffusion term even in liquid phase. This is written as,

$$\left[\dot{m}'' Y_{F,s} - \rho_s \left(D_{F,m} \frac{dY_F}{dx}\right)_s\right]_g = \left[\dot{m}'' Y_{F,s} - \rho_s \left(D_{12} \frac{dY_F}{dx}\right)_s\right]_l. \tag{4.20}$$

Here, $Y_{F,s}$ is the mass fraction of the fuel at the interface, $D_{F,m}$ is the diffusion coefficient of fuel to mixture and the partial derivative is with respect to the direction perpendicular (x-direction) to the pool surface, portraying mass diffusion due to the concentration gradient in x-direction. On the right-hand side, the second term is the diffusion of fuel in the liquid phase. Here, D_{12} is liquid-phase binary diffusivity (1.26×10^{-9} m^2/s) [3]. For non-polar liquid fuel, having one component, the second term in right-hand side can be removed and $[Y_{F,s}]_l = 1$ if the fuel has no impurities. Note that $[Y_{F,s}]_g$ on the left-hand side term of Eq. (4.20) is less than 1. For a multi-component liquid fuel, conservation of individual fuel component is considered. If the number of liquid components exceeds two, the binary diffusivity, D_{12}, is changed to diffusion coefficient as in the gas phase. Conservation of water (in the case of alcohol pool) or any other fuel component (as in a multi-component fuel), written in terms of its mass fraction Y_W, is expressed as

$$\left[\dot{m}''Y_{W,s} - \rho_s\left(D_{W,m}\frac{dY_W}{dx}\right)_s\right]_g = \left[\dot{m}''Y_{W,s} - \rho_s\left(D_{12}\frac{dY_W}{dx}\right)_s\right]_l. \quad (4.21)$$

For other species (ith species) in gas phase, the conservation is written considering the fact that species will be transported only till the interface and cannot penetrate into the liquid phase. This is expressed as,

$$\left[\dot{m}''Y_{i,s} - \rho_s\left(D_{i,m}\frac{dY_i}{dx}\right)_s\right]_g = 0. \quad (4.22)$$

The energy conservation at the interface, considering a binary liquid (for example, alcohol fuel + water) is written as,

$$\left[\sum_{j=1}^{2}\dot{m}''Y_{j,s}L_j - \sum_{j=1}^{2}\rho_s\left(D_{12}L_j\frac{dY_j}{dx}\right)_s\right]_l - \left[k_s\left(\frac{dT}{dx}\right)_s\right]_g + \left[k_s\left(\frac{dT}{dx}\right)_s\right]_l = 0. \quad (4.23)$$

Here, L_j is the latent heat of vaporization of jth species in liquid phase. The last two terms on the left-hand side of the energy conservation represent heat conduction from gas phase to the interface and from interface to the liquid phase, respectively. It should be noted here that for high boiling point liquids, heat addition from gas phase to the interface due to radiation is also considered. Additionally, the temperature of gas phase is equal to that of liquid phase at the interface. For predicting thermodynamic equilibrium, under low pressures, Raoult's law is used. Using the activity coefficient γ, which is calculated using empirical correlations, the phase equilibrium is represented as

$$X_{i,g} = \gamma_i \frac{p_i}{p} X_{i,l}. \quad (4.24)$$

Here, X_i is the mole fraction and p_i is the partial pressure of the ith species. The partial pressure is calculated as the saturation pressure at the surface temperature using equations such as Wagner's equation or Clausius-Clapeyron equation.

In general, the liquid temperature plays an important role in controlling the flame spread rate. Various regimes are observed at different liquid temperatures as shown in Fig. 4.3, for methanol, which is unique. In general, results from several experiments have shown that flame spread can be classified into regimes such as liquid controlled, gas-phase controlled and asymptotic flame spread, based on the liquid temperature. In liquid controlled regime, the flame spread rate is quite low and can be pulsating. In the gas-phase controlled regime, the flame spread rate increases to value much higher than that in the liquid controlled regime and it increases as the liquid temperature increases. In the asymptotic regime, the flame spread rate remains almost a constant and becomes insensitive to the liquid temperature.

The depth of the pool plays an important role on flame spread rate, next to the fuel type and its initial temperature. Experiments have shown that based on the type of liquid fuel, flame will not spread over its surface if the depth is below a critical value. For instance, for an n-decane pool, if the depth is less than around 2 mm, the flame will not spread. As the pool depth increases, the flame spread rate increases and asymptotically reaches a constant value. Similarly, the fuel width also plays a notable role in determining the fire spread rate up to a certain value. Also, if the temperature is more than the flash point the effect of width becomes insignificant.

The flame spread rate forms a useful data to quantify a developing fire. The area of fire (A_f) can be written in terms of the radius (r) of the flame spread (considering a circular pool ignited at its center) as,

$$A_f(t) = \pi r(t)^2. \tag{4.25}$$

If the flame spread velocity is V_f, then $r(t)$ can be written as $V_f \times t$, at any given time instant, t. Thus,

$$A_f(t) = \pi (V_f \times t)^2. \tag{4.26}$$

The heat release rate from the fire is expressed in terms of mass loss rate and heat of combustion, as,

$$\dot{Q}(t) = \dot{m}'' A_f(t) \Delta H_c. \tag{4.27}$$

This can be written using the expression for A_f as,

$$\dot{Q}(t) = \left(\dot{m}'' \pi V_f^2 \Delta H_c\right) \times t^2 = \alpha \times t^2, \tag{4.28}$$

where α is a factor dependent on characteristics of the fuel, which are estimated from flammability tests.

4.4 Mass Burning Rate

When a flame spreads over a liquid fuel surface, it primarily heats up the liquid surface downstream to evaporate more vapors out of it and continue its propagation. After the flame propagates over the entire fuel surface, based on the thickness of the fuel pool or the fuel availability, the flame consumes the fuel. For thick fuel pools, the rate at which the fuel is consumed is almost steady and the fuel surface attains an equilibrium temperature. Combustion is controlled by convection and diffusion processes. Oxygen from the oxidizer stream is transported by convection as well as by mass diffusion towards the flame zone. Fuel vapor from the pool surface is transported towards the flame zone predominantly by diffusion. Heat transfer from the flame supports continued evaporation of the fuel from the interface. Chemical kinetics plays a less important role.

As shown in Eq. (4.3), the gradient of the fuel concentration at the interface mainly determines the mass loss rate. Under convective (laminar as well as turbulent) conditions, as the fuel vapor formed over the surface is transported to the flame zone at a faster rate, the mass burning rate increases. For fires under normal gravity conditions, buoyancy flow induced due to density gradients influence the burning rate.

The mass burning rate is expressed in terms of fuel surface regression rate ($\dot{r}$), expressed in m/s, which is the rate of descent of the fuel surface in the vertical direction as the fuel burns.

$$\dot{m} = A \times \dot{r} \times \rho_l. \tag{4.29}$$

Here, A is the area of the pool and ρ_l is the density of the fuel. The mass burning rate is almost steady if the fuel thickness (thickness of the spill) is higher or if the fuel is continuously supplied at the rate at which it burns. In literature [5], regression rates of most of the petroleum fuels such as gasoline, diesel, tractor oil, and kerosene are available considering pool diameters in the range of 0.4–2300 cm. It has been observed that for pool diameter in the range of 0.4 cm–approximately 5 cm, the regression rate decreases as the pool diameter is increased. For pool diameter in the range of 5–12 cm, the regression rate remains almost a constant. The flames over the liquid pool are phenomenally laminar in these ranges of pool diameter. When the pool diameter is increased above 12 cm, the flame transitions from laminar to turbulent regime. Here, the regression rate increases with increasing pool diameter up to 1 m. The fire becomes turbulent at this stage and the regression rate reaches an almost constant value of 4 mm/min. The ratio of average flame height to the pan diameter is observed to decrease continuously in both laminar and transition regimes and it attains an almost constant value in the turbulent regime. For smaller pools, conduction from flame, walls of the pan and substrate, and convection form the primary heat transfer modes to the pool. The flame stand-off from the surface dictates the conduction heat transfer. Formation of fuel-rich vapor between the flame and the fuel surface decreases the regression rate. However, as transition to turbulence occurs, higher rates of mixing are accomplished and the regression rates increase. For pool diameter greater than 1 m, radiative heat transfer from the fire to the pool becomes dominant and the regression rate reaches a constant value irrespective of a further increase in the pool diameter. The asymptotic value of the regression rate ($\dot{r}_{max}$) for pools larger than 1 m, can be calculated as,

$$\dot{r}_{max} = C_1 \times \Delta h_c / \Delta h_v. \tag{4.30}$$

Here, C_1 is an empirical constant (1.27×10^{-6} m/s), Δh_c is the heat of combustion and Δh_v is net heat of vaporization, which is evaluated as,

$$\Delta h_v = h_{fg} + \int C_{p,l} dT. \tag{4.31}$$

Here, h_{fg} is the latent heat of vaporization at the boiling point of the liquid and the second term in the right-hand side is the sensible heat added to the liquid phase in order to increase its temperature from its initial value to the boiling point. The mass burning rate per unit area of the pool ($\dot{m}''$) increases as the pool diameter is increased

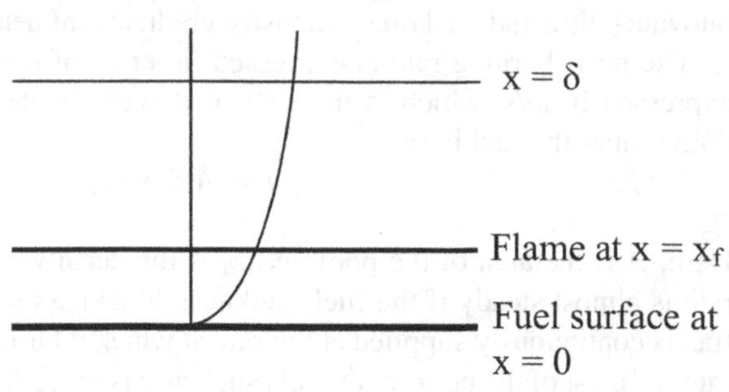

Fig. 4.4 Combustion in an almost stagnant layer

and reaches an asymptotic (maximum) value. The asymptotic mass burning rate per unit area ($\dot{m}''_{max}$) can be evaluated using the asymptotic regression rate.

$$\dot{m}''_{max} = \dot{r}_{max} \times \rho_l = C_2 \times \Delta h_c / \Delta h_v. \tag{4.32}$$

Here, C_2 is an empirical constant (10^{-3} kg/m^{2}s). Zabetakis and Burgess [6] developed an expression for mass burning rate per unit area including the pool diameter as follows:

$$\dot{m}'' = \dot{m}''_{max} \times [1 - \exp(-\kappa \beta D)]. \tag{4.33}$$

Here, κ is termed as extinction coefficient ($1/m$), β is the mean beam length correction (unit less) and D is the pool diameter (m), which is the hydraulic diameter for non-circular pools.

The height of a liquid fire depends on the heat release rate, which is a function of the mass burning rate. This is estimated by empirical correlations. Since fire is a turbulent phenomenon, the fire height is also correlated as a given percentage of its intermittency. For example, 50% intermittent flame represents a vertical location above the pool surface where the flame is observed for at least 50% of the given duration of burning. Typical flame height correlation as presented by Heskestad [7] is,

$$L_f = 0.23 \times (\dot{m} \Delta h_c)^{2/5} - 1.02 \times D, \tag{4.34}$$

where L_f is in m, $\dot{m} \Delta h_c$ in kW, and D is in m.

A theoretical analysis of steady mass burning rate is carried out using the approach related to the analysis of a diffusion flame. In such approach, only the gas-phase transport is considered. Boundary conditions at the fuel surface are written as coupling conditions between the liquid and gas phases. Considering the fact that the chemical time is much lesser than the time involved in the transport of reactant species towards the flame zone, an infinitely fast rate chemistry formulation is considered. Since the dynamics are involved in the direction normal to the fuel surface, a one-dimensional (x) coordinate in that direction is considered as shown in Fig. 4.4.

Here, the oxidizer stream is transported from the ambient towards the flame sheet located at $x = x_f$. The velocity in the direction parallel to the fuel surface is low and falls in the laminar regime. Fuel is transported to the flame sheet from the liquid surface located at $x = 0$. Both fuel and oxidizer are consumed at the flame sheet; that is, the mass fractions of fuel and oxidizer at the flame sheet are zero. Reaction zone occurs well within the boundary layer of thickness, δ, as shown in Fig. 4.4. The boundary layer of thickness, δ, is calculated as a function of Reynolds number. This formulation can be applied for burning of moderately sized liquid pools. Here, radiative heat transfer is also neglected.

The equations governing this problem are written as follows. The mass conservation in terms of mixture density and x-direction velocity is written as,

$$\frac{d}{dx}(\rho u) = 0 \Rightarrow \rho u = \dot{m}'' = \text{constant}. \tag{4.35}$$

Fuel conservation is written in terms of mass fraction of the fuel (Y_F) and the net volumetric rate (kg/m^{3}s) at which the fuel is consumed ($\dot{\omega}_f'''$).

$$\dot{m}''\frac{dY_F}{dx} = \rho D\frac{d^2Y_F}{dx^2} - \dot{\omega}_F'''. \tag{4.36}$$

Similarly, the conservation equation of the oxidizer is written as,

$$\dot{m}''\frac{dY_O}{dx} = \rho D\frac{d^2Y_O}{dx^2} - \dot{\omega}_O'''. \tag{4.37}$$

The energy conservation with chemical energy source term, representing the heat release rate in the flame ($\dot{\omega}_F'''\Delta h_c$) is written as,

$$\dot{m}''C_p\frac{dT}{dx} = k\frac{d^2T}{dx^2} + \dot{\omega}_F'''\Delta h_c. \tag{4.38}$$

Consider a global reaction, 1 kg fuel + s kg oxidizer $\rightarrow$ $(1 + s)$ kg products. If a variable, $b_{FO} = (Y_F - Y_O/s)$, is defined, it is noted that $(\dot{\omega}_F''') - (\dot{\omega}_O''')/s = 0$. This implies the governing equation involving b_{FO} does not have the non-linear source term. Such a variable is called a conserved scalar. In this analysis, properties of all species are assumed constant and Lewis number is assumed to be unity.

Similarly, the non-linear source terms in their governing equations disappear by defining variables such as:

$$b_{OT} = C_pT + \Delta h_c \times Y_O/s, \; b_{FT} = C_pT + \Delta h_c \times Y_F.$$

To solve the transport controlled combustion problem, a conserved scalar, b, defined as above, is considered. In this analysis the properties such as thermal conductivity (k), specific heat (C_p), density (ρ), and mass diffusivity (D), are considered as constants and are evaluated at a given average temperature. A constant Z is defined as $1/(\rho D)$ or C_p/k. The governing equation for the conserved scalar is then written as,

$$\dot{m}''Z\frac{db}{dx} = \frac{d^2b}{dx^2}. \tag{4.39}$$

Under steady burning conditions, $\dot{m}''$ is constant and Z is a constant involving properties calculated at an average temperature. Integrating equation (4.39),

$$\dot{m}''Zb + c_1 = \frac{db}{dx}. \tag{4.40}$$

Separating the variables in Eq. (4.40) results in,

$$dx = \frac{db}{\dot{m}''Zb + c_1}. \tag{4.41}$$

Integration of Eq. (4.41) yields,

$$\dot{m}''Z(x + c_2) = \ln(\dot{m}''Zb + c_1). \tag{4.42}$$

$$b = \frac{e^{\dot{m}''Z(x+c_2)} - c_1}{\dot{m}''Z}. \tag{4.43}$$

For the inner region between the fuel surface and the flame, the boundary conditions are: At the fuel surface ($x = 0$), $b = b_s$. At the flame, ($x = x_f$), $b = b_f$.

Invoking these in Eq. (4.43), the constants are evaluated. Finally, the solution for the inner region is given as,

$$b(x) = \frac{e^{\dot{m}''Z(x-x_f)}(b_f - b_s) - b_f e^{-\dot{m}''Zx_f} + b_s}{1 - e^{-\dot{m}''Zx_f}}. \tag{4.44}$$

For the outer region between the flame and oxidizer stream, the boundary conditions are: At the flame, ($x = x_f$), $b = b_f$. At $x = \delta$, $b = b_\infty$. Invoking these in Eq. (4.43), the solution in the outer region is expressed as,

$$b(x) = (b_\infty - b_f)\left(\frac{e^{\dot{m}''Zx} - e^{\dot{m}''Zx_f}}{e^{\dot{m}''Z\delta} - e^{\dot{m}''Zx_f}}\right) + b_f. \tag{4.45}$$

Three "b" variables involving Y_F, Y_O, and T are formulated as conserved scalars. The solution of these variables in the inner and the outer regions are presented in Eqs. (4.44) and (4.45). The boundary conditions for the primitive variables Y_F, Y_O, and T are presented in Table 4.1. The boundary conditions for the primitive variables are used to determine the boundary conditions for the "b" variables. It may be noted that the definitions of "b" variables have been modified to include constants

Table 4.1 Boundary conditions for Y_F, Y_O, and T

x	Y_F	Y_O	T
0	$Y_F = Y_{F,s}$ $\dot{m}'' = \dot{m}''Y_{F,s} - \rho D \frac{dY_F}{dx}\|_{x=0}$	0	$T = T_s$ $k\frac{dT}{dx}\|_{x=0} = \dot{m}''h_{fg}$
$\mathbf{x_f}$	0	0	$T = T_f$ $k\frac{dT}{dx}\|_{in} = k\frac{dT}{dx}\|_{out} + \dot{m}''\Delta h_c$
$\mathbf{x_\delta}$	0	$Y_O = Y_{O,\infty}$ $\frac{dY_O}{dx}\|_{x=\delta} = 0$	$T = T_\infty$ $\frac{dT}{dx}\|_{x=\delta} = 0$

Table 4.2 Definitions for "b" variable and associated boundary conditions

x	b_{FO}	b_{FT}	b_{OT}
Definition	$\frac{Y_F - Y_O/s}{Y_{F,s} - 1}$	$\frac{C_pT + \Delta h_c Y_F}{h_{fg} + \Delta h_c(Y_{F,s} - 1)}$	$\frac{C_pT + \Delta h_c Y_O/s}{h_{fg}}$
x = 0 $\frac{db}{dx} = \dot{m}''Z$	$\frac{Y_{F,s}}{Y_{F,s} - 1}$	$\frac{C_pT_s + \Delta h_c Y_{F,s}}{h_{fg} + \Delta h_c(Y_{F,s} - 1)}$	$\frac{C_pT_s}{h_{fg}}$
x = x_f	0	$\frac{C_pT_f}{h_{fg} + \Delta h_c(Y_{F,s} - 1)}$	$\frac{C_pT_f}{h_{fg}}$
x = δ $\frac{db}{dx} = 0$	$\frac{-Y_O/s}{Y_{F,s} - 1}$	$\frac{C_pT_\infty}{h_{fg} + \Delta h_c(Y_{F,s} - 1)}$	$\frac{C_pT_\infty + \Delta h_c Y_{O,\infty}/s}{h_{fg}}$

in their denominators. This is done to either normalize the variable or to make it non-dimensional.

Table 4.2 presents the "b" variables and the boundary conditions associated with them. These boundary conditions can be used to determine the unknowns T_f, $\dot{m}''$, and x_f. It should be noted that T_s and therefore, $Y_{F,s}$, are also unknowns. However, they are related using thermodynamic equilibrium at the interface, as mentioned before. For example, the equilibrium temperature when steady burning occurs can be determined using the boundary conditions and the Clausius-Clapeyron equation given as,

$$X_{F,s} = \exp((-h_{fg}/R) \times [1/T_s - 1/T_b]).$$

Here, R is the universal gas constant and T_b is the boiling point of the fuel. From this approach, choosing appropriate B-number from Table 4.2 and associated boundary conditions, for steady burning of a liquid pool/ film, Eqs. (4.44) and (4.45) can be solved and the expression for mass loss rate is given as,

$$\dot{m}'' = \frac{1}{Z\delta}\ln(1 + B). \tag{4.46}$$

For example, by choosing $b = b_{OT}$, the Spalding's transfer number is expressed as,

$$B = \frac{C_P(T_\infty - T_s) + Y_{O,\infty} \times \Delta h_c/s}{h_{fg}}. \tag{4.47}$$

The flame temperature and its location are given by the following expressions:

$$T_f = T_s + \frac{(sB - Y_{O,\infty})}{(s + Y_{O,\infty})}\frac{h_{fg}}{C_P}. \tag{4.48}$$

$$\frac{x_f}{\delta} = 1 - \frac{\ln(1 + Y_{O,\infty}/s)}{ln(1 + B)}. \tag{4.49}$$

The mass fraction of the fuel at the interface is expressed as,

$$Y_{F,s} = \frac{B - Y_{O,\infty}/s}{B + 1}. \tag{4.50}$$

4.5 Case Studies

4.5.1 Case Study 1: Steady Evaporation of Liquid Fuels

For the problem shown in Fig. 4.2, the equilibrium temperature at the liquid surface and mass evaporation rate per unit area for three liquid fuels at various ambient temperatures have been calculated equating Eqs. (4.7) and (4.17), which is the Spalding transfer number based on mass transfer and heat transfer, respectively. Figure 4.5 shows the variation of equilibrium temperature attained during steady evaporation of methanol, n-Heptane, and m-Xylene. An appropriate equation for thermodynamic equilibrium is used to arrive at mole fraction, and therefore, mass fraction of the fuel vapor at the interface, corresponding to the surface temperature. This is done until the Eq. (4.7) equals Eq. (4.17). The mass fraction of the fuel vapor at a longer distance from interface is taken as zero in the case study. The enthalpy of vaporization, normal boiling point, and molecular weight of these fuels are reported in Table 4.3.

The gas-phase properties such as ρ, D_{AB}, k, and C_p are calculated as the corresponding properties of standard air at average temperatures, following the procedure reported in Rangwala et al. [8]. The evaporation takes place at atmospheric pressure. It is clear from Fig. 4.5 that as the ambient temperature is increased, the equilibrium temperature increases towards the boiling point. However, the temperature of the liquid surface is always less than the boiling point. Steady mass evaporation rate per unit area has been calculated by considering the value of L in Eqs. (4.7) and (4.17) as 1 m.

The equilibrium surface temperature, latent heat of vaporization, and air properties at average temperatures have been used in this calculation. Figure 4.6 shows the variation of steady evaporation rate per unit area as a function of ambient temperature for methanol, n-Heptane, and m-Xylene.

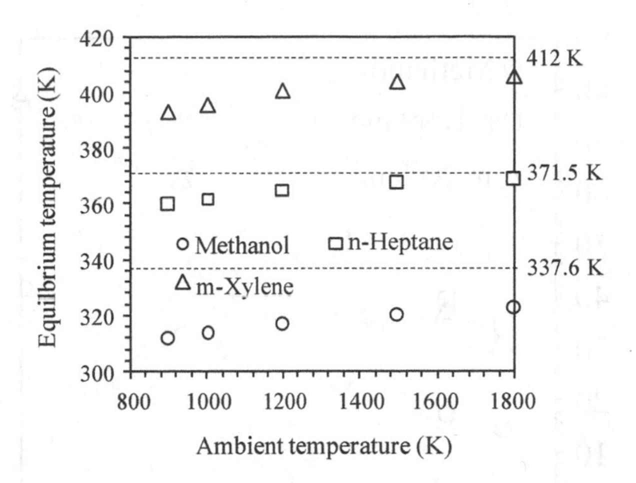

Fig. 4.5 Variation of equilibrium temperature at the liquid surface as a function of ambient temperature for methanol, n-Heptane, and m-Xylene, compared to the boiling points

Table 4.3 Thermo-physical properties of fuels

Fuel	Molecular weight (kg/kmol)	Enthalpy of vaporization h_{fg} (J/kg)	Boiling point T_b (K)
Methanol	32	1101000	337.6
n-Heptane	100	317700	371.5
m-Xylene	106	336415	412

It is seen that the evaporation flux increases with an increase in the ambient temperature as a result of an increase in the equilibrium surface temperature and the corresponding increase in the mass fraction of the fuel vapor at the interface. Due to much higher enthalpy of vaporization of methanol, its evaporation rate is lower than the other two fuels, even though its boiling point is much lower. Similar mass evaporation fluxes are observed for n-Heptane and m-Xylene.

4.5.2 Case Study 2: Steady Burning of Liquid Fuels—Numerical Simulation

Here, an opposed-flow liquid fuel burner is considered. The liquid pool is kept in a circular cylindrical burner of diameter 20 mm and an inverted co-axial cylindrical duct with same diameter that supplies air is kept on the top at a distance of 15 mm as shown in Fig. 4.7a. The liquid pool level is assumed to be maintained constant by supplying the fuel at the rate at which it burns. Methanol, m-Xylene, and Methyl Methacrylate (MMA) have been used as test liquid fuels. MMA is the liquid monomer of Poly

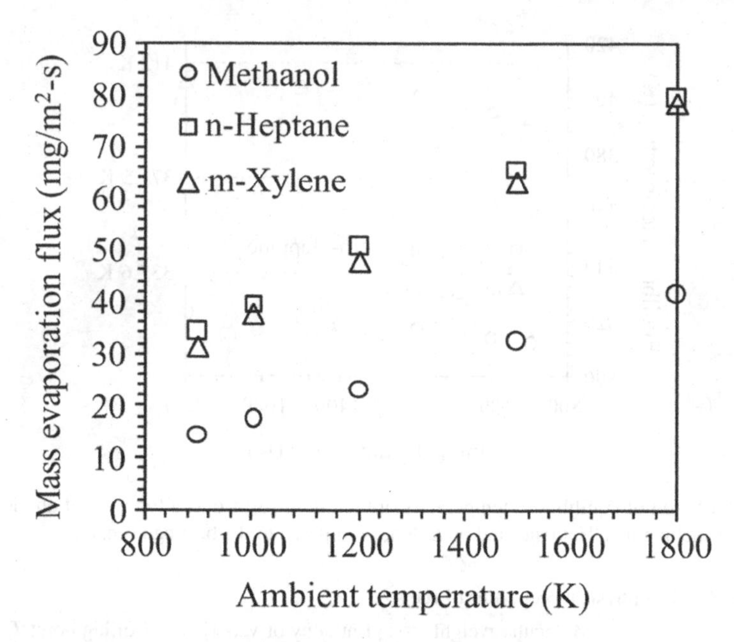

Fig. 4.6 Variation of steady mass evaporation flux as a function of ambient temperature for methanol, *n*-Heptane, and *m*-Xylene

Methyl Methacrylate, PMMA, a widely used plastic material. Numerical simulations of the burning of pool flames under opposed air flow have been carried out using interface boundary conditions as listed in Table 4.1. Ansys Fluent is used to solve the steady-state conservation equations of mass, momentum, energy, and species, in the gas-phase alone. A short chemical kinetics mechanism having 49 species and 376 reactions have been used to model the reaction kinetics.

As the air velocity is increased, the steady burning rate of the liquid fuel is seen to increase as shown in Fig. 4.7b. This is because of the faster transport of oxidizer towards the flame, which enables more fuel to burn.

Also, the higher air velocity increases the momentum of the air stream, which in turn pushes the flame towards the pool enhancing the heat transport to the fuel. It should be noted here that methanol is a non-sooty fuel and the other two are sooty. As observed in the evaporation rate, due to its higher latent heat of vaporization, methanol's burning rate is lower than other fuels, especially at higher air velocities.

The mass burning rates of *m*-Xylene are higher than MMA at all air velocities considered. The structure of the opposed flame over the *m*-Xylene pool surface is presented subsequently. Here, an air velocity of 0.4 m/s is considered. The numerically predicted profiles of temperature and mass fractions of reactant and major product species are shown in Fig. 4.8.

The flame zone is observed around 2.5 mm from the pool surface denoted by the maximum temperature and maximum mass fraction of CO_2 at that location. Fuel and oxidizer are almost consumed around this location. It is noted that both CO and H_2O

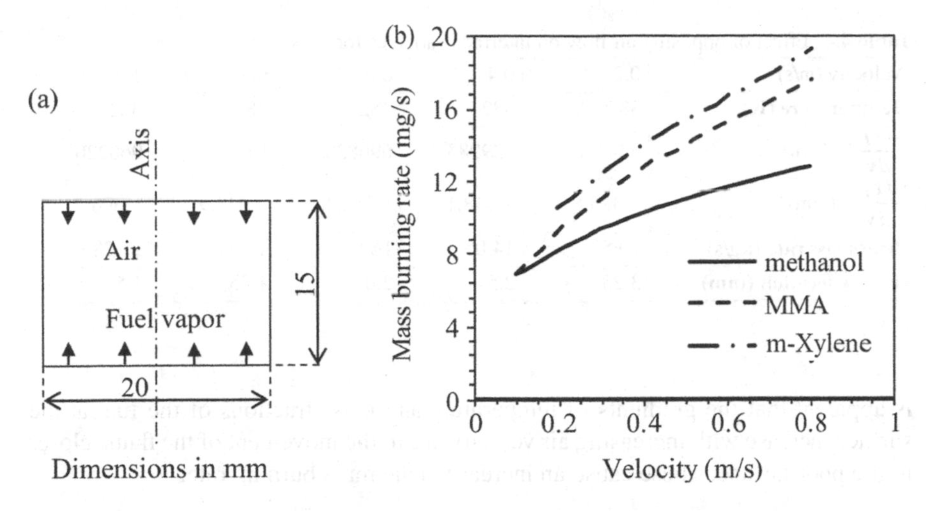

Fig. 4.7 **a** Computational domain for the opposed-flow liquid fuel burner, **b** variation of mass burning rate with opposing air velocity for liquid pools fueled by *m*-Xylene, methanol, and MMA

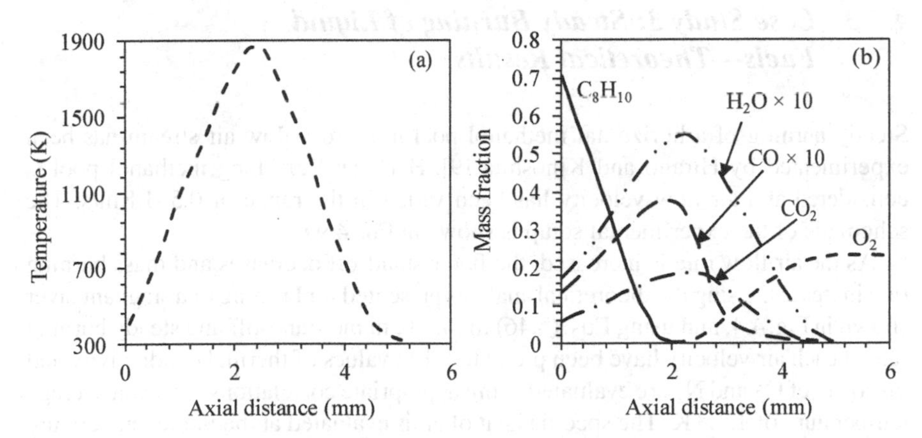

Fig. 4.8 Variation of **a** temperature and **b** mass fractions of reactants and major product species along the axis during the steady burning of *m*-Xylene pool flame for the case of air velocity of 0.4 m/s

peaks just to the left of CO_2. CO assists in the formation of both CO_2 and H_2O, in the outer and inner zones across the flame. It is evident that during the steady burning, the surface temperature has reached 382 K. The normal boiling point of *m*-Xylene is 412 K. The mass fraction of the fuel at the surface corresponding to the surface temperature of 382 K is 0.72.

In fact, irrespective of the air velocity, the surface temperature reaches an equilibrium value of 382 K as shown in Table 4.4.

Also shown in Table 4.4 are the flame locations, mass burning rates, and gas-phase side gradients of temperature and mass fraction of the fuel at the surface. It

Table 4.4 Effect of opposing air flow on interface variables for m-Xylene

Velocity (m/s)	0.2	0.4	0.6	0.8	1
Temperature (K)	382	382	382	382	382
$\frac{dT_F}{dy}\|_s$ **(K/m)**	324128	529983	696830	837214	960220
$\frac{dY_F}{dy}\|_s$ **(1/m)**	−325.8	−448.1	−546.1	−628.9	−696.7
Mass loss rate (kg/s)	10.52	14.05	16.86	19.24	21.28
Flame location (mm)	3.25	2.5	2.0	1.75	1.5

is apparent that the gradients of temperature and mass fractions of the fuel at the surface increase with increasing air velocity due to the movement of the flame closer to the pool surface. These cause an increase in the mass burning rate.

4.5.3 Case Study 3: Steady Burning of Liquid Fuels—Theoretical Results

Steady burning of a horizontal methanol pool in a cross-flow air stream has been experimented by Hirano and Kinoshita [9]. Here, an 8 cm long methanol pool is considered and air flow velocity has been varied in the range of 0.5–1.8 m/s. The schematic of the experimental setup is shown in Fig. 4.9a.

As the air flow rate is increased, the flame stand-off decreases and mass burning rate increases. Using the theoretical analysis presented for burning of a stagnant layer shown in Fig. 4.4, and using Eqs. (4.46) to (4.50), flame stand-off and steady burning rate at each air velocity have been predicted. The values of thermal conductivity and viscosity of O_2 and N_2 are evaluated using appropriate correlations [3], at an average temperature of 1323 K. The specific heat of air is evaluated at the flame temperature (calculated iteratively as 1939 K) following Rangwala et al. [8]. Mixture property for air is calculated using Wilke's mixture rule [3]. The equations for B-number [Eq. (4.47)] and flame temperature [Eq. (4.48)] are solved iteratively with specific heat value calculated at T_f using a piece-wise polynomial considering air. The values are found to be 2.72 and 1983.8 K, respectively. The maximum flame temperature measured in experiments is 1905 K. The flame stand-off is calculated by using the B-number and boundary layer thickness, is calculated using Reynolds number based on freestream conditions. Density is evaluated at ambient temperature of 300 K. The diffusion coefficient is calculated based on the assumption of $Le = 1$.

It is clear that using air standard properties at appropriate temperatures [8], theoretical relations are able to predict the flame stand-off quite closely as observed in the experiments. This is shown in Fig. 4.9b–d for air velocities of 0.5, 1 and 1.8 m/s, respectively.

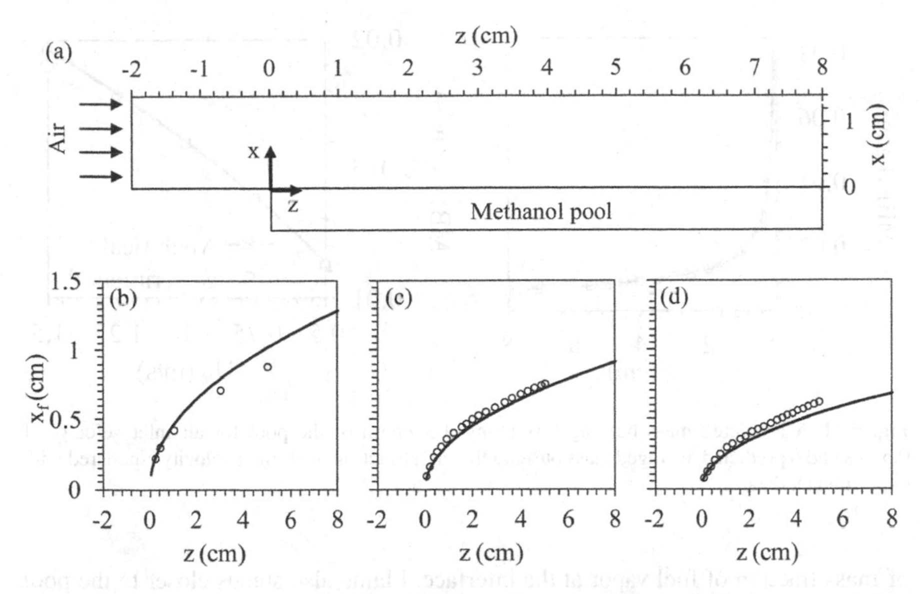

Fig. 4.9 **a** Experimental schematic, theoretically predicted flame stand-off (line) as compared to experimental data (open circles) for air inlet velocities **b** 0.5 m/s **c** 1 m/s and **d** 1.8 m/s

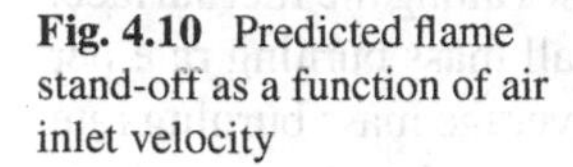
Fig. 4.10 Predicted flame stand-off as a function of air inlet velocity

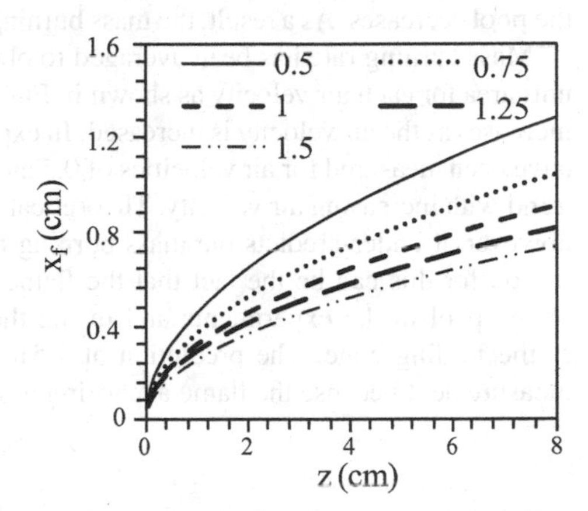

The theoretically predicted flame stand-off distances are shown in Fig. 4.10 for various air velocities, where a decrease in flame stand-off is seen with increasing air velocity. It should be noted that at an air velocity of around 1.5 m/s, the leading edge of the flame moves beyond the leading edge of the methanol pool ($z = 0$). However, the theory will not be able to predict this, because of the assumptions involved in the formulation. Mass burning rate per unit area has been plotted as a function of z for the case with air velocity of 0.5 m/s in Fig. 4.11a.

Prediction shows that the mass burning rate is high at the leading edge due to the availability of fresh oxygen from the incoming air as well as the higher gradient

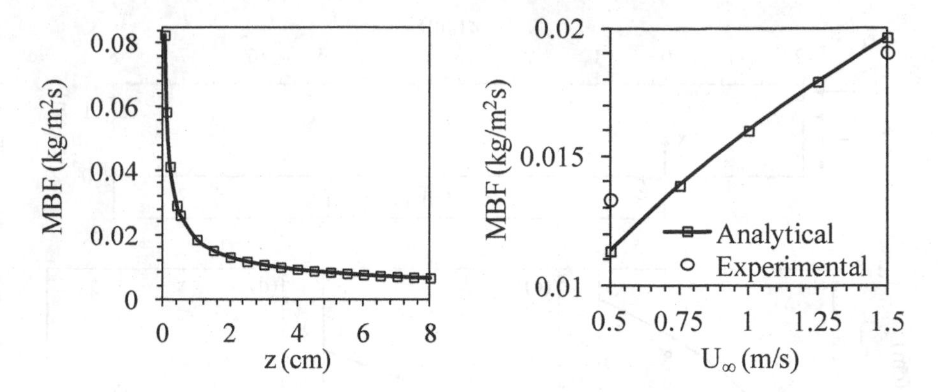

Fig. 4.11 **a** Predicted mass burning flux along the length of the pool for air inlet velocity of 0.5 m/s and **b** predicted averaged mass burning flux as a function of air inlet velocity compared with experimental data

of mass fraction of fuel vapor at the interface. Flame also stands closer to the pool surface at the leading edge of the pool. As z increases, air stream is diluted with combustion products, flame stand-off increases, and fuel mass fraction gradient at the pool decreases. As a result, the mass burning rate decreases along the fuel surface.

Mass burning rate has been averaged to obtain the overall mass burning rate per unit area for each air velocity as shown in Fig. 4.11b. The average mass burning rate increases as the air velocity is increased. In experiments, the mass burning rates (net) have been measured for air velocities of 0.5 and 1.5 m/s, and they show an increasing trend with increasing air velocity. Theoretical prediction also shows the same trend; however, it under-predicts the mass burning rate at an air velocity of 0.5 m/s. The reason for this can be the fact that the flame stands upstream of the leading edge of the pool in the experiments and in the theory it is considered to be anchored at the leading edge. The prediction at 1.5 m/s slightly overpredicts experimental measurement because the flame anchoring is slightly upstream as discussed earlier.

4.5.4 Case Study 4: Crude Oil Spill

A pipeline carrying crude oil leaks in a remote area. By the time the leak is spotted and stopped, the spill spreads over a $(10 \times 10)\ \mathrm{m}^2$ area. One way of cleaning up the oil spill quickly is by burning it. Such a process is called in-situ burning and is often used by first responders to clean up hazardous spills quickly, rather than letting them linger in the ecosystem, thereby causing long-term contamination. This case study will evaluate the criteria involved in burning of the crude oil for two spill thickness of 2 cm (Case 1) and 1 cm (Case 2). The properties of crude oil are given in Table 4.5.

Our objective is to estimate duration of the burn. We do this by using energy balance from Wu et al. [10],

Table 4.5 Properties of crude oil

Properties	Unit	Value
Liquid density	kg/m^3	868.6
Viscosity	centi-poise, cP	11.04
Fire point	°C	177
Boiling Point	°C	200
Thermal Conductivity	W/m-K	0.7
Specific Heat	kJ/kg K	2.3
Latent Heat of vaporization	kJ/kg	250
Molecular weight	kg/kmol	100

Table 4.6 Variation of oil spill thickness and regression rate with time

τ (cm)	1	2	3	4	5
y_s (cm)	0.2827	0.5118	1.2523	2.1584	3.1112
r (mm/min)	0.2354	5.7491	9.5282	10.6312	11.1019
$\dot{m}''$ (kg/m²s)	0.0034	0.0832	0.1379	0.1539	0.1607

$$\dot{q}_s'' = H_v \times \rho_F \times r(t) + \dot{q}_c'',$$

where $\dot{q}_s''$ is the heat flux from the flame on the surface of the oil, H_v is the latent heat of vaporization, and ρ_F is the fuel liquid density. $r(t)$ is the regression rate of the fuel surface given by $r(t) = \frac{\partial y_s(t)}{\partial t}$, where $y_s(t)$ is the oil spill thickness. It is to be noted that initially at $t = 0$ s, $y_s(0) = 2$ cm (case 1), and $y_s(t) = 1$ cm (case 2). $\dot{q}_s''$ is given by Wu et al. [10] as,

$$\dot{q}_s'' = \chi \frac{4\rho_\infty C_{P,\infty}}{\pi} \left[T_\infty g(T_f - T_\infty)\right]^{1/2} d^{1/2},$$

where, $\chi = 0.005$, ρ_∞ is the ambient air density, $C_{P,\infty}$ is the specific heat of ambient air, T_f is the flame temperature (1100 K), and d is the flame diameter. The in-depth conduction term $\dot{q}_c''$, is equal to

$$\frac{k(T_s - T_\infty)}{y_s(t)},$$

where k is the thermal conductivity of the oil, and T_s is the surface temperature which can be approximated as the fire point of the fuel.

For a confined pool, the flame diameter equals the equivalent diameter calculated from spill area equal to using $10 \times 10\ m^2$. However, if the spill is unconfined which is normally the case, the maximum fire size is calculated using $A_f = 1.55 \times A_s$ [10]. Since $A_f = 1.55 \times 100 = 155\ m^2 = \frac{\pi d^2}{4}$, flame diameter $d = 14.05$ m. Assuming that the crude oil has also weathered because of which the lighter volatiles evaporate,

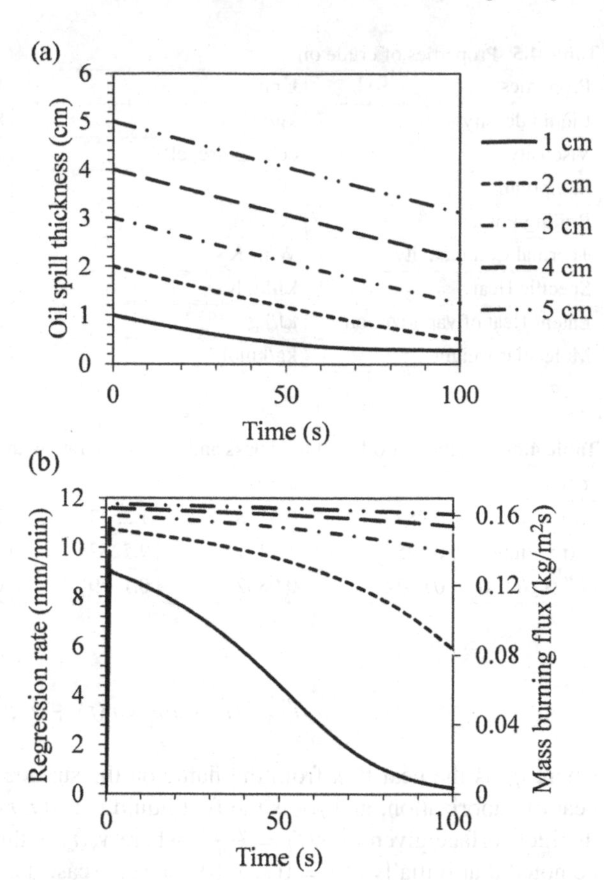

Fig. 4.12 **a** Predicted time evolution of oil spill thickness, and **b** predicted regression rate and mass burning flux for crude oil spill of thickness 1 and 2 cm

its flash point and fire point increase. Using T_s to be fire point (450 K), $T_\infty = 298$ K, $\rho_\infty = 1.2\,\text{kg/m}^3$, $C_{P,\infty} = 1005$ J/kg-K, and $T_f = 1100$ K in the energy conservation equation, we get

$$\frac{dy_s(t)}{dt} = 2.03 \times 10^{-4} - \frac{5.71 \times 10^{-7}}{y_s(t)}$$

The ordinary differential equation can be solved using RK4 method. For 1 and 2 cm thick oil spill, $y_s(t)$ is obtained as shown in Fig. 4.12a. Other initial thickness 3, 4 and 5 cm are also shown. Further, the regression rate as

$$r(t) = \frac{dy(t)}{dt}$$

and mass burning flux as

$$\dot{m}'' = \rho_F \times r(t)$$

can be determined as shown in Fig. 4.12b. These values are also tabulated in Table 4.6.

Review Questions

1. What are the important properties associated with a liquid fuel?
2. What are the significances of the flash and the fire points?
3. Fuels A and B with flash point of $-43\,°C$ and $39\,°C$, respectively. Classify the fuels based on the flash point temperatures and suggest which fuel could be considered for application: (a) in jet engines, (b) as a pilot fuel.
4. What are the factors contributing to the driving force for liquid vaporization?
5. Write the significance of the B-number for evaporation.
6. Consider evaporation of a 4 mm deep liquid contained in a 10 mm long tube in quiescent air. The saturation temperature of the liquid is 335 K. The liquid surface temperature reaches 10 K below the saturation temperature in an ambient environment of 400 K. The specific heat, thermal conductivity, and latent heat of vaporization of the liquid are 2.53 kJ/kg-K, 0.2 W/m-K, and 600 kJ/kg, respectively. Assuming equi-diffusion of the liquid (Le = 1), evaluate: (a) Spalding transfer number, (b) evaporation rate, (c) vapor mass fraction and temperature at a distance 2 mm away from the liquid interface.
7. An n-Heptane liquid pool initially kept at ambient conditions (P = 1 atm, T = 298 K) is heated to a temperature 10 K below the boiling point (Boiling point = 371.42 K). The concentration of fuel becomes zero at a height of 10 cm above the pool surface. Estimate the vapor mass fraction at the interface, the Spalding's transfer number, mass flux of fuel, and the expression for fuel mass fraction profile. Take the binary diffusivity and enthalpy of vaporization of n-Heptane—air as $0.78 \times 10^{-5}\ m^2/s$ and 317.7 kJ/kg, respectively. (Use Clausius-Clapeyron equation for calculating the interface mass fraction).
8. A dodecane liquid pool kept at 400 K and 1 atm pressure evaporates into ambient air (Boiling point = 489.2 K). If the binary diffusivity of dodecane—air is $0.81 \times 10^{-5}\ m^2/s$ obtain the expression for mass flux of dodecane. The pool is heated and the surface temperature attains 480 K. If the concentration of fuel becomes zero at 15 cm above the liquid surface, calculate the mass flux of fuel at 5 cm above the pool. Take the enthalpy of vaporization as 303.5 kJ/kg.
9. Explain the term phase equilibrium.
10. A flame spreads over a rectangular pool burner of dimensions 4 cm × 2 cm at the rate of 0.23 cm/s. If the overall heat lost to the ambient is 2.7 kW, determine the minimum burner power rating in kW in order to sustain the combustion for 1 min.
11. How does the pool diameter affect the mass burning rate?
12. A diffusion flame is formed over MMA pool of diameter 150 mm. If the surface regresses at a rate of 1 mm/min, estimate the height of the flame. The heat of combustion and density of MMA can be taken as 25527 kJ/kg and $940\ kg/m^3$, respectively.

13. Explain the factors affecting the flame spread rate over a liquid.
14. What is a conserved scalar?
15. Write the "*b*" variables associated with steady burning of a liquid pool.
16. Write about the factors that affect the burning rate of a liquid fuel.
17. An oil spill in offshore drilling area spreads for a distance of 1.5 km (diameter). If the thickness of the oil film is 5 mm and the mass burning rate is 12 kg/s, estimate the time required for removing the spill and determine the surface regression rate. The density and calorific value of the fuel can be taken as 750 kg/m^3 and 43000 kJ/kg, respectively.
18. Consider a steady cross-flow flame formed over a 4 cm long *n*-Hexane pool surface when air at 320 K enters at free stream velocities of 0.5, 1, and 1.5 m/s. Assuming the flame temperature to be 2270 K and evaluating the air and fuel properties at the specific temperatures, determine the mass loss rate, the fuel mass fraction at the interface, and plot the flame stand-off along the fuel length, for the different velocities.
19. Ambient air at 300 K is blown over a methanol pool at a velocity of 0.3 m/s. The pool is ignited and a boundary layer flame sustains. If the thickness of the boundary layer is 1 mm and the surface temperature of the pool is 5 K below the boiling point (Boiling point = 337.7 K), evaluate the mass loss rate of the fuel, the temperature and location of the flame and the mass fraction of fuel at the interface. The properties (specific heat, thermal conductivity, density) may be evaluated for air at an average temperature of 1200 K. Assume Le = 1 in the calculations. The latent heat of vaporization and heat of combustion of methanol are 1100 and 22343 kJ/kg, respectively.

References

1. Gottuk, D.T., and D.A. White, Liquid fuel fires. *SFPE handbook of fire protection engineering*, 3rd ed. Section 2, 297–317.
2. Glassman, I., R.A. Yetter, and N.G. Glumac. 2015. *Combustion*. Elsevier.
3. Reid, R.C., J.M. Prausnitz, and B.E. Poling. 1977. *The properties of gases and liquids*. Tata McGraw Hill.
4. Akita, K. 1973. Some problems of flame spread along a liquid surface. *Proceedings of the Combustion Institute* 14 (1): 1075–1083.
5. Blinov, V.I., and G.N. Khudiakov. 1961. Diffusion burning of liquids. *U.S. Army Translation*, NTIS No. AD296762.
6. Zabetakis, M.G., and D.S. Burgess. 1961. Research on hazards associated with the production of and handling of liquid hydrogen. *U.S. Bureau of Mines Report*. RI 5705.
7. Heskestad, G. 1983. Luminous heights and turbulent diffusion flames. *Fire Safety Journal* 5: 103–108.
8. Rangwala, A.S., V. Raghavan, J. Sipe, and T. Okano. 2009. A new property evaluation scheme for mass transfer analysis in fire problems. *Fire Safety Journal* 44 (4): 652–658.
9. Hirano, T., and M. Kinoshita. 1974. Gas velocity and temperature profiles of diffusion flame stabilized in stream over liquid fuel. *Proceedings of the Combustion Institute* 15: 379–387.

10. Wu, N., M. Baker, G. Kolb, and J.L. Torero. 1996. Ignition, flame spread and mass burning characteristics of liquid fuels on a water bed. *Spill Science and Technology Bulletin* 3 (4): 209–212.

Chapter 5
Burning of Solids

The chapter will focus on four areas covering solid fuels fires:

1. Solid-phase thermal decomposition or pyrolysis and conditions for sustained flame or ignition.
2. Heat and mass transfer to the unburned fuel or flame spread.
3. The heat released from the combustion processes or burning rate.
4. Conditions for no-flame-spread or suppression and no flame/extinction.

An understanding of these phenomena is essential from a perspective of evaluating fire risk. A safe separation between buildings, for example, is an application of pyrolysis and ignition. An overall assessment of the fire load and consequent impact on the structure is an application of burning rate and flame spread. The quantity of water needed to control a given size fire is an application of no-flame-spread (suppression) and no flame (extinction).

5.1 Pyrolysis and Ignition

Ignition is defined as observing a sustained diffusion flame in the volatile gases formed from a solid exposed to a heat source. The molecules comprising the solid involve thousands of atoms in cyclic bonds and cannot be directly oxidized quickly. Because of this reason, when exposed to heat, the molecules tend to break apart into smaller molecules, which can ignite by mixing with air. This process of molecule breakdown is called pyrolysis, and the released fuel vapor is called pyrolyzate. For example, in other fields of research, like Chemistry and Material Science, pyrolysis is the degradation in the absence of oxygen. In Fire Science, however, pyrolysis is taken to be in the presence of air. It should be noted that pyrolysis is an irreversible process. Thus, the solid heating will produce vapors which on condensation will not

A. S. Rangwala and V. Raghavan, *Mechanism of Fires*,
https://doi.org/10.1007/978-3-030-75498-3_5

produce the original solid. This is not the case for the ignition of pure liquids, for example, when the liquid-vapor has the same chemical composition of the liquid. In thermodynamic terms, the vaporization of the liquid is reversible. Another distinction worth noting is that the heating of liquid sets up convective motion in the liquid.

In most practical fire problems, typical heat sources are (i) convective heating from hot gases or flames; (ii) thermal radiation from the surrounding flames, (iii) conduction from a hot element on the surface (hot ember, for example). From the perspective of fire, ignition is important because it initiates fire and is responsible for fire growth. The spread of fire between objects or from room to room in a building is an ignition process. This is related to the time it takes from the commencement of heating to the onset of sustained flaming. Thus, the study of ignition and flame spread boils down to determine (theoretically or experimentally) the controlling parameters to determine this time. Following Torero [1], the time to ignition is controlled by environmental parameters (external control) as

1. Temperature, pressure, and oxygen concentration of flow; and velocity of the gas around the solid.
2. The magnitude, uniformity, and spectral quality of the incident radiation.
3. Sample geometry, thickness, and size of the exposed area.
4. Location and size of the pilot.
5. The orientation of the surface exposed to the heat flux as this influences the buoyancy and corresponding flow around the heated object.

In addition, internal factors also control the time to ignition:

1. Thermo-physical and thermochemical properties of the solid.
2. Moisture content.
3. Radiation properties such as spectral absorptivity and transmissivity.
4. Spectral transmittance of the pyrolysis gases released by the surface due to heating.
5. Chemistry of gas-phase combustion.
6. Influence of melting and bubble formation bursts on the surface because of increasing gas pressure inside the solid.
7. Kinetics of the thermal solid-phase decomposition chemistry and composition of decomposition products, including char formation rate for charring materials.

Table 5.1 shows a list of critical quantities used in the past to define ignition criteria based on theoretical or experimental analysis of ignition. These quantities form a subset of the environmental and internal parameters discussed above, which were most suitable for developing theory or setting up the experiment. The ignition criteria listed in Table 5.1 depend on the problem that needs to be solved and available data from the literature, calculations, or a model.

To solve the ignition problem using any of the criteria from Table 5.1 necessitates a fundamental understanding of three areas:

Table 5.1 Ignition criteria in fire problems (adapted from Annamalai and Durbetaki [2])

Criterion	Medium	Description
Thermal explosion or bulk explosion theory	Gases	Semenov's thermal theory [3]
Explosion theory	Gases	Chemical explosion by the buildup of radicals [4]
Clausius-Clapeyron thermal equilibrium condition	Liquid	Thermodynamic equilibrium [5]
Ignition temperature	Solids/liquids	The temperature at the location of interest exceeds a specific value [6]
Critical evolution	Solids/liquids	The temperature at which a pyrolyzing material liberates the pyrolyzate at a specified rate [7]
Critical heat release	Solids/liquids	The temperature at which the total or local heat release rate exceeds a specified value [4]
Thermal gradient	Solids/liquids	Local temperature gradient exceeds a specified value [4]
Inflexion	Solids/liquids	The rapid rate of change in the pyrolyzate mass fraction at the interface [4]

1. Heat and mass transfer associated with external heating of a surface.
2. Chemistry of solid-phase decomposition.
3. Gas-phase runaway reactions leading to the onset of ignition.

In most fire problems of practical interest, only area 1: heat and mass transfer associated with external heating of a surface is considered, and ignition models are correlated with a simple thermal model based on inert heating of the solid (one or some variation of ignition criteria's 4–6 in Table 5.1). This is because the time taken for gas-phase reactions is very fast compared to solid heating time and thus can be ignored. For example, Quintiere [3] estimates the time for gas-phase reactions to be $\sim 2 \times 10^{-4}$ s, gas-phase transport, or mixing of fuel vapor with air $\sim$4 s, and time for inert heating of solid (pyrolysis time) $\sim$120 s. The pyrolysis time is an order of magnitude higher than the chemical or mixing time. At the same time, there are certain unique instances [1] where mixing times and pyrolysis times are comparable, and both need to be accounted for. However, in most practical fire problems, they can be ignored. Additional information on why the above assumptions are reasonable is discussed in Torero [1].

Figure 5.1 shows photographs of piloted ignition of a solid fuel sample imposed with a radiative heat flux from above. Figure 5.2 shows the surface temperature during the ignition of red oak under a similar experimental setup where the red oak sample is exposed to a radiative heat flux from above.

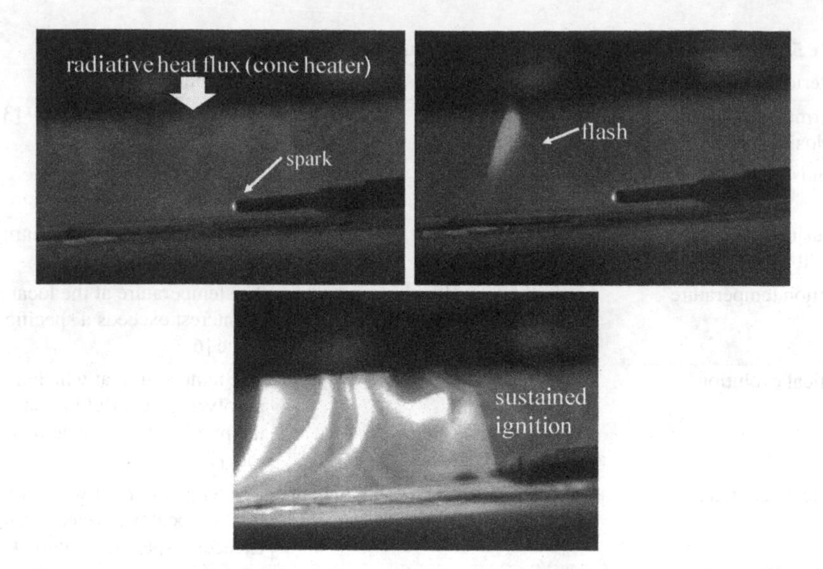

Fig. 5.1 Process of ignition of a solid fuel sample with imposed radiative heat flux

After the commencement of external heating, temperatures throughout the solid gradually increase, being highest at the surface. This temperature is shown plotted with time in Fig. 5.2 [8]. As the surface of the solid temperature increases, combustible volatiles form around the solid surface and mix with the surrounding air in the mass transfer boundary layer to produce a fuel-air mixture. Because of the presence of a continuous spark, when the volatile rate is sufficient to form a pocket of flammable gas, it tends to flash. The flash is a sudden increase in temperature at the surface corresponding to a flash of flame as shown in Fig. 5.1. The flash is not sustained and immediately extinguishes. The flame is momentary in the form of a flash because it cannot generate enough heat to overcome the heat losses on the solid surface. The fuel surface needs to be externally heated further, such that its surface temperature increases.

Figure 5.2 shows that the red oak sample flashes three times until the surface's temperature reaches a critical temperature, whereby the flash finally leads to sustained flaming. For red oak, the measured critical temperature equals 395 °C as shown in Fig. 5.2. At this surface temperature, the red oak sample can release enough fuel vapor (pyrolyzate) such that the flash results in a sustained flame. The solid line shown in Fig. 5.2 represents the calculated temperature rise of red oak assuming it is an inert solid using the equation $\rho_s C_s \frac{\Delta T}{\Delta t} = \dot{q}_e''$, where ρ_s = density of red oak (500 kg/m^3), C_s = specific heat of red oak (1380 J/kg K), and $\dot{q}_e''$ = external heat flux (18.8 kW/m^2). Thus, if the red oak sample were assumed to be inert, then the onset of ignition would be predicted by a temperature of 372 °C.

The above example also demonstrates that for solid fuel ignition, it is necessary for the surface temperature caused by external heating to rise to some critical value

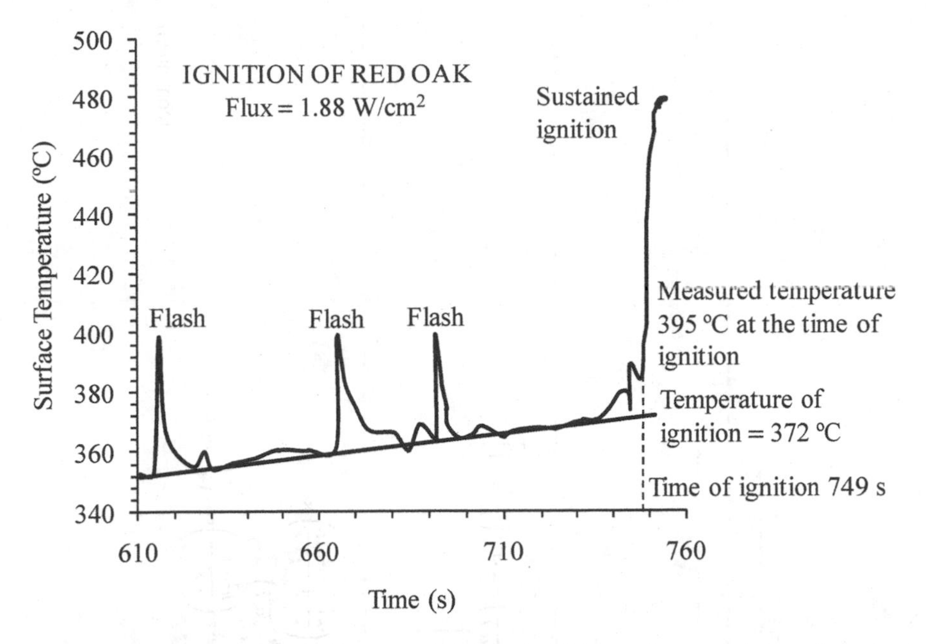

Fig. 5.2 Surface temperature profile versus time for a Red Oak Sample exposed to a radiative heat flux of 1.88 W/cm^2. Data from Atreya and Wichman [8]

such that a critical mass flux is generated. This was first suggested by Roberts and Quince [9] for a liquid fuel and extended to a solid fuel by Rasbash and Drysdale [10]. In this condition, for a gas-phase reaction to occur, the theoretical flame temperature is around 1300 °C for most hydrocarbons [3]. The flame temperature increases with decreasing oxygen concentration, as shown by Quintiere and Rangwala [11].

Modeling Ignition in Fire Ignition of a gas-air mixture in a pilot's presence is a combustion process whereby a non-reacting fuel-air mixture suddenly becomes reactive and leads to the liberation of its exothermic potentiality. In liquids and solids, the ignition is more complicated as we also need to form the gas-air mixture. This involves the solution of the vaporization in liquids and pyrolysis in the case of solids. Both involve an energy balance at the fuel-air interface. In Fire, ignition is modeled, assuming the liquid or solid behaves as an inert upto the point of ignition.

The time to ignition and temperature of the surface at ignition are the two quantities of interest which can be calculated by solving the heat equation given by:

$$\rho_s C_s \frac{\partial T}{\partial t} = k_s \frac{\partial^2 T}{\partial x^2}, \tag{5.1}$$

where ρ_s is the solid's density, C_s is the specific heat of the solid, and k_s is the solid's thermal conductivity. The equation is a linear second-order parabolic partial differential equation and can be solved analytically for certain initial and boundary conditions. Table 5.2 shows the equation's solutions for all initial and boundary condi-

Table 5.2 Analytical Solutions to Ignition Problems

Initial and BC [T(x,t)]	Schematic	Solution
Semi infinite solid		
Constant Surface Temperature $T(x, 0) = T_i$ $T(0, t > 0) = T_s$ (Nellis and Klein [13])	T_s, x	$\frac{T(x,t) - T_i}{T_s - T_i} = 1 - \text{erf}\left(\frac{x}{2\sqrt{\alpha_s t}}\right)$ $\dot{q}''_{x=0} = \frac{k_s}{\sqrt{\pi \alpha_s t}}(T_s - T_i)$
Constant Surface Heat Flux $T(x, 0) = T_i$ $-k_s \left.\frac{\partial T}{\partial x}\right\|_{x=0} = \dot{q}''_s$ (Nellis and Klein [13])	$\dot{q}''_s$, x	$T(x,t) - T_i = \frac{\dot{q}''_s}{k_s}\left[2\sqrt{\frac{\alpha_s t}{\pi}} \exp\left(-\frac{x^2}{4\alpha_s t}\right) - x \,\text{erfc}\left(\frac{x}{2\sqrt{\alpha_s t}}\right)\right]$ $T_{ig} = T(0, t_{ig}) = T_i + \frac{\dot{q}''_s}{k_s}\left[2\sqrt{\frac{\alpha_s t_{ig}}{\pi}}\right]$ $t_{ig} = \left(\frac{k_s}{\dot{q}''_s}\frac{(T_{ig} - T_i)}{2\sqrt{\alpha_s/\pi}}\right)^2$
Constant Convective Heating $T(x, 0) = T_i$ $-k_s \left.\frac{\partial T}{\partial x}\right\|_{x=0} = h(T_\infty - T_{x=0})$ (Nellis and Klein [13])	h, T_∞, x	$\frac{T(x,t) - T_i}{T_\infty - T_i} = \text{erfc}\left(\frac{x}{2\sqrt{\alpha_s t}}\right) - \exp\left(\frac{hx}{k_s} + \frac{h^2 \alpha_s t}{k_s^2}\right) \times$ $\text{erfc}\left(\frac{x}{2\sqrt{\alpha_s t}} + \frac{h}{k_s}\sqrt{\alpha_s t}\right)$ $T_{ig} = T(0, t_{ig}) = T_\infty - (T_\infty - T_i)\left[\exp\left(\frac{h^2 \alpha_s t_{ig}}{k_s^2}\right)\text{erfc}\left(\frac{h}{k_s}\sqrt{\alpha_s t_{ig}}\right)\right]$

(continued)

Table 5.2 (continued)

Initial and BC [T(x,t)]	Schematic	Solution
Constant Surface Heat Flux and Convective Cooling $T(x,0) = T_i$ $-k_s \frac{\partial T}{\partial x}\Big\vert_{x=0} = h(T_\infty - T_{x=0}) + \dot{q}_s''$ (Quintiere [14])	$\dot{q}_s''$, $h,\ T_\infty$, $\dot{q}_{re}''$, x	$T(x,t) - T_i = \frac{\dot{q}_s''}{h_T}\left[\text{erfc}\left(\frac{x}{2\sqrt{\alpha_s t}}\right)\right] - \frac{\dot{q}_s''}{h_T}\exp\left(\frac{h_T x}{k_s} + \frac{h_T^2 \alpha_s t}{k_s^2}\right)\text{erfc}\left(\frac{x}{2\sqrt{\alpha_s t}} + \frac{h_T\sqrt{\alpha_s t}}{k_s}\right)$ $T_{ig} = T(0,t_{ig}) = T_i + \frac{\dot{q}_s''}{h_T}\left[1 - \exp\left(\frac{h_T^2 \alpha_s t_{ig}}{k_s^2}\right)\text{erfc}\left(\frac{h_T}{k_s}\sqrt{\alpha_s t_{ig}}\right)\right]$ $h_T = h + \frac{\epsilon\sigma\left(T(0,t)^4 - T_\infty^4\right)}{T(0,t) - T_\infty}$
Integral model		
Constant Surface Heat Flux and Convective Cooling $T(x,0) = T_i$ $-k_s \frac{\partial T}{\partial x}\Big\vert_{x=0} = h(T_\infty - T_{x=0}) + (1-\epsilon)\dot{q}_s''$ (Atreya [12])	$\dot{q}_s''$, $h,\ T_\infty$, $\varepsilon\sigma\left(T_s^4 - T_\infty^4\right)$, x	$\theta^*(x,t) = \theta_s^{*2}(t^*)\left[1 - x^*\left(\frac{\Phi_s}{\theta_s^*}\right) + x^{*2}\left(\frac{\Phi_s^2}{4\theta_s^{*2}}\right)\right]$ $\frac{t^*}{\theta_s^*} = \frac{1}{3}\left(\frac{1}{\Phi_s^2} - \frac{10AB}{\theta_s^*(A^2 - 4B)} - 2\frac{2(A + 2B\theta_s^*)}{\theta_s^*(A^2 - 4B)\Phi_s}\right)$ $M_s^* = \frac{A^*(1-\rho_f^*)}{\Phi_s E^*}(\theta_s^* + 1)^2 \exp\frac{-E^*}{\theta_s^* + 1}\left[1 - \frac{\exp\frac{-E^*\theta_s^*}{\theta_s^*+1}}{(\theta_s^* + 1)^2}\right]$ $M_s^* = \frac{C_s h^*(\theta_f^* - \theta_s^*)Y_{O\infty}/\nu C_p}{\Delta H - \theta_f^* - (\theta_f^* - \theta_s^*)Y_{O\infty}/\nu}$
$A = -(h^* + 4\sigma^*),\ B = \frac{-25\sigma^*}{3},\ \theta^* = \frac{T_s - T_\infty}{T_\infty},\ \rho^* = \frac{\rho_s}{\rho_\infty},\ \rho_f^* = \frac{\rho_{sf}}{\rho_{s\infty}},\ E^* = \frac{E}{RT_\infty},\ h^* = \frac{hT_\infty}{\epsilon\dot{q}_s''},\ t^* = t\left[\frac{k_s/(\rho_s C_s)}{l^2}\right]$		

Table 5.2 (continued)

Initial and BC [T(x,t)]	Schematic	Solution
Surface Energy Pulse per Unit Area $T(x,0) = T_i$ $\lim_{\Delta t \to 0} \dot{q}'' \Delta t = E''$ $\left.\frac{\partial T}{\partial x}\right\|_{x=0} (t>0) = 0$ (Nellis and Klein [13])	E" x	$T(x,t) - T_i = \frac{E''}{\rho_s C_s \sqrt{\pi \alpha_s t}} \exp\left(-\frac{x^2}{4\alpha_s t}\right)$ $T_{ig} = T(0, t_{ig}) = T_i + \frac{E''}{\rho_s C_s \sqrt{\pi \alpha_s t_{ig}}}$ $t_{ig} = \left[\frac{E''}{\rho_s C_s (T_{ig} - T_i)\sqrt{\pi \alpha_s}}\right]^2$
Sinusoidal Surface Temperature $T(x,0) = T_i + \Delta T sin(\omega t)$ (Nellis and Klein [13])	T (t) x	$T(0,t) = T_i + \Delta T \sin(\omega t)$ $T(x,t) - T_i = \Delta T \exp\left(-x\sqrt{\frac{\omega}{2\alpha_s}}\right) \sin\left(\omega t - x\sqrt{\frac{\omega}{2\alpha_s}}\right)$ $\dot{q}''_{x=0} = k_s \Delta T \sqrt{\frac{\omega}{\alpha_s}} \sin\left(\omega t + \frac{\pi}{4}\right)$
Contact between two Solids $T(x,0) = T_{i,A},\ x < 0$ $T(x,0) = T_{i,B},\ x > 0$ (Nellis and Klein [13])	$T_{i,A}$ $T_{i,B}$ A B x_A 0 x_B	$T_{i,0} = T(0,0) = \frac{\sqrt{k_A \rho_A C_A}\, T_{i,A} + \sqrt{k_B \rho_B C_B}\, T_{i,B}}{\sqrt{k_A \rho_A C_A} + \sqrt{k_B \rho_B C_B}}$ $\frac{T_A(x,t) - T_{i,A}}{T_{i,0} - T_{i,A}} = 1 - \mathrm{erf}\left(\frac{x_A}{2\sqrt{\alpha_A t}}\right)$ $\frac{T_B(x,t) - T_{i,B}}{T_{i,0} - T_{i,B}} = 1 - \mathrm{erf}\left(\frac{x_B}{2\sqrt{\alpha_B t}}\right)$

Table 5.2 (continued)

Thermally thin solid (Bi < 0.1)		
Initial and BC [T(t)]	Schematic	Solution
$T(0)=T_i$ $\rho_s C_s \tau \frac{dT}{dt} = h[T_\infty - T(t)] + a\dot{q}_s''$ Babrauskas [4]	$\dot{q}_s''$, h, T_∞, $(1-a)\dot{q}_s''$, τ	$T(t)-T_i = \frac{a\dot{q}_s''}{h}\left[1-\exp\left(\frac{-ht}{\rho_s C_s \tau}\right)\right]$ $t_{ig} = \frac{\rho_s C_s \tau}{h}\ln\left(\frac{a\dot{q}_s''}{\dot{q}_s'' - h(T_{ig}-T_i)}\right)$
$T(0)=T_i$ $\rho_s C_s \tau \frac{dT}{dt} = 2h[T_\infty - T(t)] + (2a-1)\dot{q}_s''$	$\dot{q}_s''$, h, T_∞, h, T_∞, $(1-a)\dot{q}_s''$, $(1-a)\dot{q}_s''$, τ	$T(t)-T_i = \frac{a\dot{q}_s''}{2h}\left[1-\exp\left(\frac{-2ht}{\rho_s C_s \tau}\right)\right]$ $t_{ig} = \frac{\rho_s C_s \tau}{2h}\ln\left(\frac{a\dot{q}_s''}{\dot{q}_s'' - 2h(T_{ig}-T_i)}\right)$
$T(0)=T_i$ $\rho_s C_s \tau \frac{dT}{dt} = 2h[T_\infty - T(t)] + 2a\dot{q}_s''$	$\dot{q}_s''$, $\dot{q}_s''$, h, T_∞, h, T_∞, $(1-a)\dot{q}_s''$, $(1-a)\dot{q}_s''$, τ	$T(t)-T_i = \frac{a\dot{q}_s''}{h}\left[1-\exp\left(\frac{-2ht}{\rho_s C_s \tau}\right)\right]$ $t_{ig} = \frac{\rho_s C_s \tau}{2h}\ln\left(\frac{a\dot{q}_s''}{\dot{q}_s'' - h(T_{ig}-T_i)}\right)$
$T(0)=T_i$ $\rho_s C_s \tau \frac{dT}{dt} = 2h[T_\infty - T(t)]$	h, T_∞, h, T_∞, τ	$\frac{T(t)-T_i}{T_\infty - T_i} = \exp\left(\frac{-2ht}{\rho_s C_s \tau}\right)$ $t_{ig} = \frac{\rho_s C_s \tau}{2h}\ln\left(\frac{T_\infty - T_i}{T_\infty - T_{ig}}\right)$

Table 5.2 (continued)

Reacting solid: iterative solution required for time to ignition $\rho_s C_s \frac{\partial T}{\partial t} = k_s \frac{\partial^2 T}{\partial x^2} + \rho_s A Q \exp\left(\frac{-E}{RT}\right)$		
$T(t=0) = T_i$ Fixed heat flux $\dot{q}_s''$ (Liñán and Williams [15])	$\dot{q}_s''$ $\dot{q}_R'''$	$T_{ig} = \frac{E}{R}\left[\ln \frac{k_s \rho_s C_s T_i^2 A}{(\dot{q}_s'')^2}\right]^{-1}$ $t_{ig} = \frac{\pi k_s \rho_s C_s T_i^2}{4\dot{q}_s''}\left[\frac{\frac{E}{RT_i}}{\ln \frac{k_s \rho_s C_s (T_{ig})^2 A}{0.65(\dot{q}_s'')^2 \left(\frac{E}{RT_i}\right)^{1/2}}\left[\frac{\pi(\dot{q}_s'')^2 t_{ig}}{k_s \rho_s C_s T_i^2}\right]^{1/4}} - 1\right]^2$
Porous solid with porosity ϵ_P $T(t=0) = T_i$ Fixed heat flux $\dot{q}_s''$ (Telengator *et al.* [16])	$\dot{q}_s''$ $\dot{q}_R'''$	$t_{ig} = \frac{\pi k_s \rho_s C_s T_i^2}{4\dot{q}_s''}\left[\frac{\frac{E}{RT_i}}{\ln \frac{k_s \rho_s C_s (T_{ig})^2 A}{0.65(\dot{q}_s'')^2 \left(\frac{E}{RT_i}\right)^{1/2}}\left[\frac{\pi(\dot{q}_s'')^2 t_{ig}}{k_s \rho_s C_s T_i^2}\right]^{1/4}} - 1\right]^2$ $\times(1-\epsilon_P)\left(1 - \epsilon_P + \frac{\lambda_g \epsilon_P}{\lambda_s}\right)$

ϵ_P = Void volume/Total volume, $\dot{q}_R'''$ = Volumetric source term

Table 5.2 (continued)

Spontaneous ignition or auto ignition of thermally thick solids

$$\rho_s C_s \frac{\partial T}{\partial t} = k_s \frac{\partial^2 T}{\partial x^2} + \rho_s A Q \exp\left(\frac{-E}{RT}\right)$$

Initial and BC	Schematic	Time to ignition
$T(t=0) = T_i$ Fixed temperature T_s on one side (Vilyunov and Zarko [17])	T_s, $\dot{q}_R'''$	$t_{ig} = \frac{RT_s^2}{E}\frac{C_s}{QA}\exp\left[\frac{E}{RT_s}\right]\left[1 + \frac{2E}{RT_s^2}(T_s - T_i) + 0.163\left(\frac{E}{RT_s^2}(T_s - T_i)\right)^2\right]$
$T(t=0) = T_i$ (Beever [18])	$\dot{q}_R'''$	$t_{ig} = M\left(\frac{RT_s^2}{E}\right)\left(\frac{C_s}{QA}\right)\exp\frac{E}{RT_i}\left[\frac{\delta}{\delta_c} - 1\right]^{1/2}, \delta = \frac{E}{R}\frac{\rho_s Q}{k_s}\frac{\tau^2}{T_R^2}A\exp\left(\frac{-E}{RT_R}\right)$ Infinite plane layer: M = 1.534, $\delta_c = 0.878$ Infinitely long cylinder: M = 1.429, $\delta_c = 2$; Sphere: M = 1.316, $\delta_c = 3.32$

Table 5.2 (continued)

Ignition of liquid: iterative solution for ignition temperature	
Clausius-Clapeyron equation: $X_{F,s} = \frac{P_{sat}}{P} = \frac{Y_{F,s} MW_{mix}}{MW_F(1 - Y_{F,s}) + Y_{F,s} MW_{mix}} = \exp\left[\frac{MW_F L}{R}\left(\frac{1}{T_{sat}} - \frac{1}{T_{ig}}\right)\right]$	$\frac{Y_{F,s}\Delta H_c + C_P(T_{ig} - T_\infty)}{(Y_{F,s} - 1)\Delta H_c + L} = \frac{Y_{F,s} + Y_{O,\infty}/s}{Y_{F,s} - 1}$
Ignition of gas: $mC_v(dT/dt) = hS(T_\infty - T) + \dot{m}'''_F V(\Delta H_c)$	
Semenov solution for auto-ignition: $T_{ig} = T_\infty\left[1 + \frac{RT_\infty}{E}\right]$	$t_{ig} = 0.632\left[\frac{C_P T_\infty}{\Delta H_c(E/RT_\infty)Ae^{-E/RT_\infty}}\right]$

Table 5.2 (continued)

Symbol	Parameter	Unit		Symbol	Parameter	Unit
a	Absorptivity of the solid	–		α	Thermal diffusivity	m^2/s
A	Pre-exponential constant	s^{-1}		δ	Frank-Kamenetskii parameter	–
C	Specific heat capacity	kJ/kg-K		ϵ	Emissivity	–
erf()	Error function	–		ϵ_P	Porosity	–
erfc()	Complementary error function	–		θ	Non-dimensional temperature	–
E	Activation energy	kJ/kmol		λ	Thermal conductivity	W/m-K
E''	Surface energy pulse	J/m^2		ν	Stoichiometric oxygen to fuel ratio	kg/kg
h	Convective heat transfer coefficient	W/m^2-K		ρ	Density	kg/m^3
h_T	Linearized heat transfer coefficient	W/m^2-K		τ	Characteristic dimension	m
ΔH_c	Heat of combustion	kJ/kg		ϕ	Net non dimensional heat flux	–
k	Thermal conductivity	W/m-K		ω	Angular velocity of sinusoidal wave	rad/s
L	Heat of vaporization at saturation temperature	kJ/kg		M	Geometrical constant	–
MW	Molecular weight	kg/kmol		P	Pressure	kPa
$\dot{q}''_s$	Incident heat flux	W/m^2		Q	Heat of reaction	J/kg
R	Universal gas constant	kJ/kmol-K		S	Surface area	m^2
T	Temperature	K		V	Volume	m^3
x	x-coordinate	m		Y	Mass fraction	–

∞ - Ambient, A - Solid A, B - Solid B, c - Critical, F - Fuel, g - Gas, i - Initial, ig - Ignition

P - Pressure, re - Re-radiation, R - reference, s - Solid/Surface/Oxygen to fuel mass ratio

sat - Saturation, $*$ - Non-dimensional

tions typically encountered in *Fire*. It should be noted that the four main assumptions used are (a) inert heating, (b) one-dimensional heating, and (c) constant properties independent of temperature, and (d) heat transfer coefficient is constant. Table 5.2 shows several cases, however, the most common formula is:

$$T(x,t) - T_i = \frac{\dot{q}_s''}{h_T}\left[\mathrm{erfc}\left(\frac{x}{2\sqrt{\alpha_s t}}\right)\right] - \frac{\dot{q}_s''}{h_T}\exp\left(\frac{h_T x}{k_s} + \frac{h_T^2 \alpha_s t}{k_s^2}\right) \times \mathrm{erfc}\left(\frac{x}{2\sqrt{\alpha_s t}} + \frac{h_T\sqrt{\alpha_s t}}{k_s}\right). \quad (5.2)$$

In Eq. (5.2), $T(x,t)$ is the temperature at x-location in the solid as a function of time. The heating is assumed to be one dimensional. $\dot{q}_s''$ is the heat flux (kW/m^2) on the surface of the solid, h_T is the total heat transfer coefficient including convective and radiative transfer, and α_s represents the thermal diffusivity of the solid = $\frac{k_s}{\rho_s C_s}$ in m^2/s. Substituting $x = 0$ in Eq. (5.2) gives the solution for the ignition temperature, T_{ig}, given by:

$$T_{ig} = T(0, t_{ig}) = T_i + \frac{\dot{q}_s''}{h_T}\left[1 - \exp\left(\frac{h_T^2 \alpha_s t_{ig}}{k_s^2}\right)\mathrm{erfc}\left(\frac{h_T\sqrt{\alpha_s t_{ig}}}{k_s}\right)\right]. \quad (5.3)$$

Equation (5.3) is the general solution to the surface temperature at all incident heat flux levels. As the solid's surface increases its temperature, it will also start radiating heat to its surroundings. This effect is accounted for by a linearized radiative heat transfer coefficient

$$h_T = h + \frac{\epsilon\sigma\left[T(0,t)^4 - T_\infty^4\right]}{T(0,t) - T_\infty}. \quad (5.4)$$

This is necessary as it is not possible to obtain an analytical solution with a boundary condition that includes radiative heat loss with temperature to the fourth power. An integral approach is needed in this case as shown by Atreya [12]. The complex form of the error function in Eq. (5.3) can be simplified, and solutions have been proposed in the literature [3, 4, 7]; however, with modern calculators, Eqs. (5.2) and (5.3) can be easily solved.

The heat equation can be simplified for cases with Bi < 0.1, where Bi is the Biot number given by $\frac{hl}{k_s}$, where h is the convective heat transfer coefficient (W/m^2-K), l is the thickness of the material (m), and k_s is the thermal conductivity of the material (W/m-K). When Bi < 0.1 holds, the lumped capacitance method can solve the transient heating during ignition. This assumption is called the thermally thin approximation. The assumption implies that the temperature of the solid is spatially uniform $\left(\frac{dT}{dx} = 0\right)$. Note that when Bi < 0.1 is true and the lumped capacitance method is used, Eq. (5.1) no longer applies as the heat equation is a differential equation governing the spatial temperature distribution, which is now assumed to be zero when the solid is thermally thin. The transient temperature response is determined by formulating an overall energy balance on the solid given by,

$$\rho_s V C_s \frac{dT}{dt} = hA_s(T_\infty - T), \tag{5.5}$$

where $\rho_s V C_s(dT/dt)$, represents the energy storage term and $hA_s(T_\infty - T)$ is the energy loss to the ambient at T_∞ by convection. Note that there is no conductivity k_s that appears in the thermally thin equation. The convective heat transfer coefficient h can be replaced by the total heat transfer coefficient h_T from Eq. (5.4). In fire, four boundary conditions arise during thermally thin ignition problems and the solutions to each are shown in Table 5.2. These conditions are

1. Front face exposed to a radiant heat flux and also undergoing convective cooling and re-radiation while back face is perfectly insulated or adiabatic. This can be approximated, for example, by a thin layer of flammable coating or a fabric on a wall.
2. Front face is exposed to a radiative heat flux and also undergoing convective cooling and re-radiation while the back face undergoes convective cooling and re-radiation. This can be approximated by a fabric or a curtain that is exposed to a heat flux from one side, for example, from an electric lamp.
3. Front and back sides are exposed to identical radiative heating and also undergo convective cooling and re-radiation. This can be approximated by a situation where the curtain is exposed to the heat flux from the electric lamp from both sides.
4. Front and back sides are exposed to convective heating. This can be approximated, for example, by a vertical curtain heating on both sides by a smoke plume.

Thermally thin materials for ignition problems in Fire are ~1 mm in thickness and hence most practical materials are usually thermally thick in nature. Another problem with the thermally thin method is the fact that the solution does not take into account fuel degradation and fuel exhaustion which can be important especially if the material is so thin. This is the reason that in some experiments, thin samples exposed to radiative heat fluxes of relatively high magnitude (~50 kW//m^2) smoke and char but do not ignite because there is not enough material to release enough pyrolyzate to sustain a Lower Flammability Limit (LFL) condition at the ignition source.

Table 5.2 shows the many combinations of radiation, convection and conduction that a solid can be subjected to under different configurations. For example, the surface of a combustible solid may be subjected to radiative heat flux, but cooled by free convection as the temperature rises. The free convection can be described by a convective heat transfer coefficient, h_c, that can be obtained from standard heat transfer textbooks. The expression in Eq. (5.3) can now be used to obtain the temperature of the surface. If the temperature reaches the characteristic ignition temperature of the fuel, then we can say that the constant radiative heat flux is sufficient to ignite the fuel.

In spite of its utility, it should be noted that the concept of an ignition temperature as a means of predicting onset of flaming combustion in solid fuels is an approximation. Thus, T_{ig} will differ based on ignition experiments, however the difference is small enough for the concept to be useful. The reason the difference is small and most problems can be solved using this concept is because the chemical rates are strongly dependent on temperature. This fact was explained earlier by using characteristic times derived by Quintiere [3], where pyrolysis time was shown to be an order of magnitude higher than chemical times. This is why we can approximate that when a surface reaches a certain temperature, namely ignition temperature, sufficient pyrolyzate vapor issues out and we can say that ignition has occurred. The basis of the existence of an ignition temperature is fundamentally the same as the basis of an existence of a fixed temperature during steady burning.

Ignition of Reacting Solids The governing equation for ignition (Eq. (5.1)) can also include a volumetric source (or sink) term, as denoted by $\dot{q}_R'''$ in Table 5.2.

$$\rho_s C_s \frac{\partial T}{\partial t} = k_s \frac{\partial^2 T}{\partial x^2} + \rho_s A Q \exp \frac{-E}{RT}, \tag{5.6}$$

where the source term is represented using an Arrhenius expression where A is the pre-exponential constant $(1/s)$ and E (kJ/mol) represents the activation energy. Q (kJ/kg) is the energy released. In this case, when the temperature of the body reaches a certain critical temperature, internal heating is triggered which can lead to smoldering or flaming combustion. Although several condensed phase reactions occur, for modeling purposes, all the reaction source and sink terms are represented by a global Arrhenius term with a global activation energy and pre-exponential constant as shown in Eq. (5.6). The solutions for different cases for self-heating are shown in Table 5.2 for porous solids with a constant heat flux. A case study (Sect. 5.5) shows a situation where use of Eq. (5.6) is made to solve a self-heating problem.

5.2 Flame Spread

Flame spread is the phenomena of a moving flame in close proximity to the source of the fuel vapor originating from the condensed phase, i.e., solid or liquid. The fuel vapor originates because of pyrolysis or vaporization of the condensed phase because of heat transfer from the advancing flame itself. The flame spread determines the fire's destiny. It applies to fire in urban conflagrations, forest fires or the first growth after ignition in a compartment fire. Figure 5.3 shows a photograph of a flame moving on a thick plastic surface with an ambient flow velocity of 0.8 m/s. The region of the fuel surface that is pyrolyzing is the pyrolysis length x_p as shown in Fig. 5.3.

The flame extends forward much beyond x_p shown by l_{h2} in Fig. 5.3. This extension heats up the unignited fuel causing it to pyrolize and ultimately ignite and the flame moves forward in the direction of the flow at a relatively high spread rate. Typical concurrent (in the direction of flow velocity) spread rates are of the order of 0.1 cm/s for condensed fuels. So if the initial ignited region were say 6 cm and the width of the plastic slab is 6 cm (area = 36 cm^2), then in 1 min the pyrolysis length will increase by 0.1 cm/s × 60 s = 6 cm. The new total pyrolysis length, x_p,

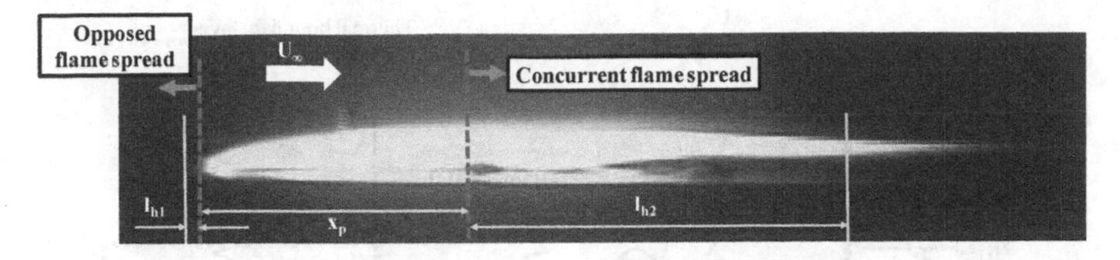

Fig. 5.3 Photograph of a flame spreading on PMMA slab (2.5 cm thick) in a flow velocity equal to 0.8 m/s. The pyrolysis length (x_p) is the region of the fuel surface where temperature equals the pyrolysis temperature of PMMA equal to 378 °C. Opposed flame spread and concurrent flame spread direction are shown by blue arrows

will now be 12 cm and the pyrolysis area would be $12 \times 6 = 72\,\text{cm}^2$. Most plastics burn with a burning rate per unit area of ~10 g/m²s and with a heat of combustion ΔH_C ~40 kJ/g, thus giving a burning rate increase from 1.4 kW to 2.8 kW in 1 min for the one-dimensional flame spread $\left(40\frac{\text{kJ}}{\text{g}} \times 10\frac{\text{g}}{\text{m}^s\text{-s}} \times 0.0036\,\text{m}^2\right)$. This makes concurrent flame spread a high hazard in fire.

The flame also propagates in the opposite direction (opposed flame spread) as shown in Fig. 5.3. In this case, the pre-heated length is shown by l_{h1} in Fig. 5.3. This length is much smaller and the resulting flame spread rate in the opposed direction is also lower. Typical values of opposed flame spread are around 0.01 cm/s, an order of magnitude lower than concurrent flame spread.

There is a similarity between the flame spread and ignition process where after a fuel has been ignited, the flame must heat up the unignited fuel ahead of the flame, and the spread of a flame can be viewed as a sequence of ignitions of a "heated length" caused by a flame that acts as both a heat source and a pilot. The relatively simple expression is the basis for quantifying the rate at which a flame spreads on a solid surface [3]:

$$\text{Flame spread rate} = \frac{\text{Flame heated length } (l_{h1} \text{ or } l_{h2})}{\text{Time to ignite } (t_{ig})}. \tag{5.7}$$

The flame heated length will depend on the nature of the flow field as shown in Fig. 5.3 where both l_{h1} and l_{h2} indicate the heating length. For opposed-flow spread, the flame heating length shown as l_{h1} is shorter. The heat flux from the flame is mostly in the form of conductive heating at the leading edge as shown in Fig. 5.4. Essentially, at this region, the flame tends to slowly propagate against the direction of the flow, and the flame spread rate can be obtained by performing an energy balance in the region shown by CV_1 in Fig. 5.4.

The energy balance around CV_1 gives the classic expression first derived by de Ris [20]: in 1969 for flame spread for a thermally thin solid opposed flame spread given by

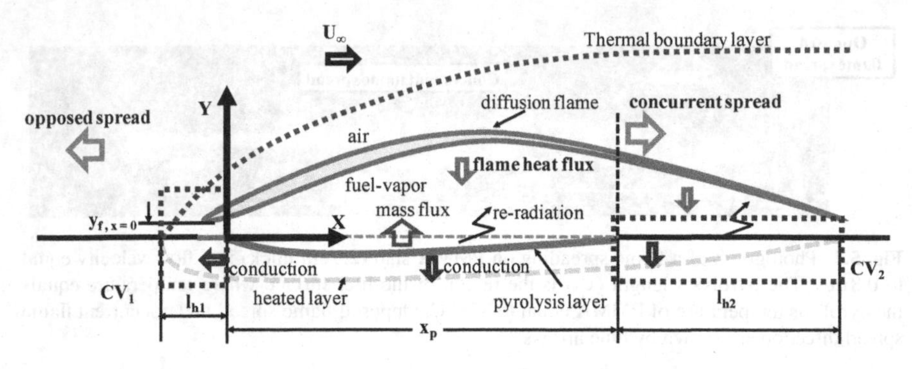

Fig. 5.4 Schematic of the flame spread problem (adapted from Pello [19])

$$V_F = \frac{\sqrt{2}k_g(T_f - T_p)}{\rho_s C_{p,s} \tau_s (T_p - T_\infty)}. \tag{5.8}$$

And opposed flame spread for a thermally thick solid is given by:

$$V_F = \frac{k_g \rho_g C_{p,g} U_\infty (T_f - T_p)^2}{k_s \rho_s C_{p,s} (T_p - T_\infty)^2}. \tag{5.9}$$

Both equations can be derived by performing an energy balance in CV_1 shown in Fig. 5.4. Assuming that the flame's leading edge heats up the fuel by conduction only, then the flame heat flux can be estimated as

$$\dot{q}_f'' = k_g \frac{T_f - T_p}{y_{f,\,x=0}}, \tag{5.10}$$

where $T_f - T_p$ is the difference between the flame temperature and the pyrolysis temperature and $y_{f,x=0}$ is distance that the flame sits above the fuel surface at the leading edge as shown in Fig. 5.4. Now, using what we learned earlier, the ignition temperature of thermally thin solid heated by an incident heat flux $\dot{q}_f''$ from the flame is given by

$$T(t) - T_i = \frac{\dot{q}_f''}{h}\left[1 - \exp\left(\frac{-ht}{\rho_s C_s \tau_s}\right)\right], \tag{5.11}$$

where the exponential term for small ignition times can be approximated by a Taylor series expansion as $\exp\left(\frac{-ht}{\rho_s C_s \tau_s}\right) \approx 1 - \frac{ht}{\rho_s C_s \tau_s}$, thus simplifying Eq. (5.11) to:

$$T(t) - T_i \approx \frac{\dot{q}_f'' t}{\rho_s C_s \tau_s}. \tag{5.12}$$

And solving for ignition time (t_{ig}) gives,

$$t_{ig} \approx \frac{\rho_s C_s \tau_s (T_p - T_\infty)}{\dot{q}_f''}. \tag{5.13}$$

The opposed-flow flame spread can now be solved by using Eq. (5.7), and substituting Eqs. (5.10) and (5.13) for the time to ignition for a thermally thin solid:

$$\text{Flame spread rate} = \frac{l_{h1}}{\text{Time to ignite } (t_{ig})} = \frac{l_{h1}}{\frac{\rho_s C_s \tau_s (T_p - T_\infty)}{\dot{q}_f''}}. \tag{5.14}$$

$$\text{Flame spread rate} = \frac{l_{h1}}{y_{f,\,x=0}} k_g \frac{T_f - T_p}{\rho_s C_s \tau_s (T_p - T_\infty)}. \tag{5.15}$$

Equation (5.15) shows the coupling between the ignition problem we studied earlier with the flame spread. Given knowledge of the time to ignition for a thermally thin solid, we see that the only unknown now is $= \frac{l_{h1}}{y_{f,\,x=0}}$, representing heated length divided by the distance of the flame from the fuel surface also called as the flame stand-off distance at the leading edge ($x = 0$). de Ris [20] showed that $l_{h1} = \sqrt{2} y_{f,\,x=0}$, thereby connecting the flame heated length and flame heat flux and substituting this expression in Eq. (5.15) gives Eq. (5.8) which is the exact solution. Similarly, Eq. (5.9) can be derived by using the same procedure described in Eqs. (5.10)–(5.14), but substituting the time to ignition for a thermally thick solid fuel $t_{ig} \approx k_s \rho_s C_s \left(\frac{T_p - T_\infty}{\dot{q}_f''}\right)^2$. In this case, the flow velocity will also influence the problem and $y_{f,\,x=0} \sim \alpha_g / U_\infty \sim l_{h1}$. If the solid fuel were aligned vertically, then the opposed-flow spread direction would be vertically downwards. In this case, the flame spread rate can be calculated by substituting the forced flow velocity U_∞, by a buoyancy induced velocity given by $U_b = \left(\frac{g \nu_g \Delta H_c Y_{O,\infty}}{C_{p,g} T_\infty}\right)^{1/3}$ [21]. For instance, with $\Delta H_c \sim 40$ kJ/g, $Y_{O,\infty} = 0.233$, and using air properties at an average temperature, U_b can be evaluated to be around 35 cm/s.

Table 5.3, adapted from Pello [19], shows the different expressions for flame spread velocity used to calculate laminar flame spread in forced and natural convection conditions for thermally thick and thermally thin fuels. In the table, $\dot{q}_e''$ is the external radiant heat flux, $\dot{q}_{fr}''$ is the radiant flux from flame to the solid, and $\dot{q}_{rs}''$ is the re-radiation from solid surface as shown in Fig. 5.4. The subscript "g" denotes gas phase and "s" denotes solid phase C_1 is an empirical constant for the solid surface heating process in a forced convective flow field. C_2 equals flame length/pyrolysis length, or x_f / x_p.

As discussed earlier, the expressions involve the time to ignition and an expression for the flame heated length. The time to ignition will change depending on thermally thick or thin condition and the flow field as discussed in Sect. 5.1 and Table 5.2. The flame heated length will depend on the flow field and direction of flame propagation

Table 5.3 Expressions for opposed and concurrent flame spread velocity, V_F (m/s) for forced and natural convective conditions

Case	Expression
Forced convection	
Opposed spread, Thermally thick	$\frac{\rho_g C_{P,g} k_g (T_f - T_p)^2 U_\infty}{\rho_s C_s k_s (T_p - T_\infty)^2}$
Opposed spread, Thermally thin	$\frac{\sqrt{2} k_g (T_f - T_p)}{\rho_s C_s (T_p - T_\infty) \delta_s}$
Concurrent spread, Thermally thick	$\frac{4 x_p C_2 \left(\left[C_1 k_g \rho_g C_{p,g} \frac{u_\infty}{x_p} \right]^{1/2} (T_f - T_p) + \dot{q}''_e + \dot{q}''_{fr} - \dot{q}''_{rs} \right)^2}{\pi k_s \rho_s C_s (T_p - T_\infty)^2}$
Concurrent spread, Thermally thin	$\frac{x_p C_2 \left(\left[C_1 k_g \rho_g C_{p,g} \frac{u_\infty}{x_p} \right]^{1/2} (T_f - T_p) + \dot{q}''_e + \dot{q}''_{fr} - \dot{q}''_{rs} \right)}{\rho_s C_s (T_p - T_\infty) \delta_s}$
Natural convection	
Opposed spread, Thermally thick	$\frac{k_g \rho_g C_{P,g} (T_f - T_p)^2 U_b}{k_s \rho_s C_s (T_p - T_\infty)^2}$
Opposed spread, Thermally thin	$\frac{\sqrt{2} k_g (T_f - T_p)}{\rho_s C_s (T_p - T_\infty) \delta_s}$
Concurrent spread, Thermally thick	$\frac{4 x_p C_2 \left(\left[\frac{g k_g \rho_g C_{P,g} (T_f - T_\infty)}{T_\infty x_p} \right]^{1/4} (T_f - T_p) + \dot{q}''_e + \dot{q}''_{fr} - \dot{q}''_{rs} \right)^2}{\pi k_s \rho_s C_s (T_p - T_\infty)^2}$
Concurrent spread, Thermally thin	$\frac{x_p C_2 \left(\left[\frac{g k_g \rho_g C_{P,g} (T_f - T_\infty)}{T_\infty x_p} \right]^{1/4} (T_f - T_p) + \dot{q}''_e + \dot{q}''_{fr} - \dot{q}''_{rs} \right)}{\rho_s C_s (T_p - T_\infty) \delta_s}$

(opposed or concurrent). The variation of opposed and concurrent flame spread, with flow velocity, under forced conditions is shown in Fig. 5.5. The flame spread rate has been shown for both thermally thick (PMMA with thickness of 1.27 cm) and thin (Whatman Chromatography paper with thickness of 0.33 mm) solids, under ambient conditions of pressure and temperature with $Y_{O,\infty} = 0.233$. Note that for the thermally thin case, both thermal conductivity, k_s, of the solid and the free stream flow velocity U_∞ do not influence the flame spread rate as shown in Table 5.3. At low flow velocities, air is entrained as the buoyant gases leave and this natural convection becomes the controlling factor as it sustains the opposed flame spread. With increasing opposed-flow velocity, the flame at the leading edge moves closer to the surface. This should cause an increase in mass and heat flux to the condensed fuel. However, at the same time the solid fuel ahead of the flame is also increasingly convectively cooled because of the increase in the convective heat transfer at the solid air interface due to increasing U_∞. These two competing heat transport mechanisms

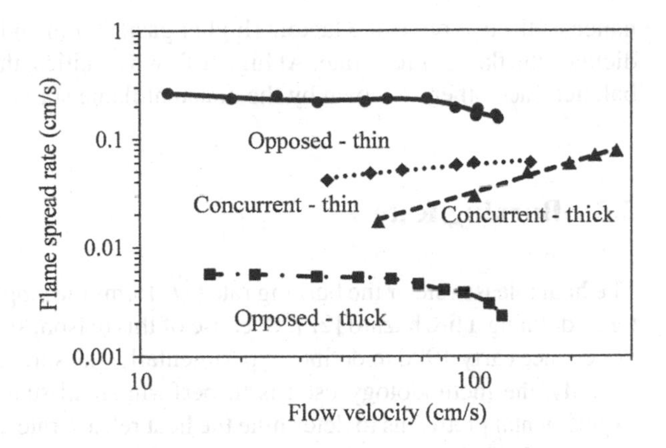

Fig. 5.5 Variation of flame spread rate with flow velocity, for opposed and concurrent flame spread. Adapted from Pello [19]. (Thick material: PMMA, Thin material: Paper)

cancel each other out making the flame spread rate independent of U_∞ for thermally thin materials. This is why the expressions for opposed flame spread under forced and natural convection are the same for the thermally thin condition, as shown in Table 5.3. This is further shown in Fig. 5.5 where the flame spread rate is almost a constant with the flow velocity for thin paper. However, for thick material such as PMMA, the flame spread is seen to reduce with flow velocity in Fig. 5.5. This is due to the dominance of the convective cooling of the solid surface ahead of the flame, leading to a reduction in the heat transfer and eventually a decrease in the flame spread rate.

Table 5.3 also shows expressions for concurrent flame spread which is much faster as it is facilitated by the direct convective and radiative heating of the flame as shown in Figs. 5.3 and 5.4. The flame heated length in this case is shown as l_{h2}. The concurrent flame spread rate can be obtained by an energy balance in the region depicted by CV_2 in Fig. 5.4. Similar to the opposed flame case, the basic idea is to determine the time it takes to heat up the surface of the solid to an ignition temperature or pyrolysis temperature by accounting for all the relevant heat gain and heat loss mechanisms. The heat fluxes ($\dot{q}''_e$, $\dot{q}''_{fr}$, and $\dot{q}''_{rs}$) involved in the problem are shown in Fig. 5.4. The pyrolysis length, x_p, is defined as the length of the solid fuel that is undergoing pyrolysis at a given instant of time, and the concurrent flame spread velocity can be expressed as $\frac{dx_p}{dt}$. The stand-off distance is smallest at the leading edge y_f (0) and increases as $x^{1/2}$ for purely forced flow and by $x^{1/4}$ for buoyant upward spread.

The concurrent flame spread over thick material shows an increasing trend with flow velocity (Fig. 5.5). As the flow velocity increases, the boundary layer thickness reduces and this enhances the heat transfer to the unburnt fuel and the flame spread rate increases.

For thin material, it is observed that the flame spread increases slightly with flow velocity (Fig. 5.5). As flow velocity increases, the rate of gasification is enhanced which leads to higher burn out and a decrease in pyrolysis length. Hence, the dominance of the two opposing factors (higher gasification and reduced pyrolysis length) dictates the flame spread rate. At higher flow velocities, these two effects are seen to balance each other as shown by the constant flame spread rate.

5.3 Burning Rate

The heat release rate or the burning rate ($\dot{Q}$) forms the single most important parameter in defining a fire hazard [22]. Because of this reason, significant research has been done since early 1900 to define, experimentally measure, and predict this parameter. Usually, the methodology used is to perform small-scale experiments in standard experimental platforms to determine the heat release rate and then use theory, empirical models, and numerical modeling to predict the behavior at larger scales typically expected in practical fire problems.

Figure 5.6 shows a turbulent diffusion flame over a condensed fuel surface (solid or liquid). The smoke plume rises up vertically. During combustion, the fuel burns, and if steady state is assumed it will slowly regress or deplete as shown by the downward arrow in Fig. 5.6.

Six different ways can be used to establish the heat release rate or burning rate or fire power or energy release rate (kW) of the process:

1. The mass of fuel consumed per unit time, $\dot{m}_{F,s}$ (g/s), multiplied by the heat of combustion ΔH_c (kJ/g) of the fuel gives the burning rate $\dot{Q}$ in kW.
2. The mass of fuel vapor, $\dot{m}_{F,g}$ (g/s), released per unit time multiplied by the heat of combustion. Note that for conservation of mass to hold, $\dot{m}_{F,s} = \dot{m}_{F,g}$.
3. The energy release is expressed in terms of kJ/g of O_2 consumed. As discussed in Chaps. 1 and 2, for many hydrocarbons this number is a constant and equal to 13.1 kJ/g. Thus, if the rate at which oxygen is consumed in the fire is known, then

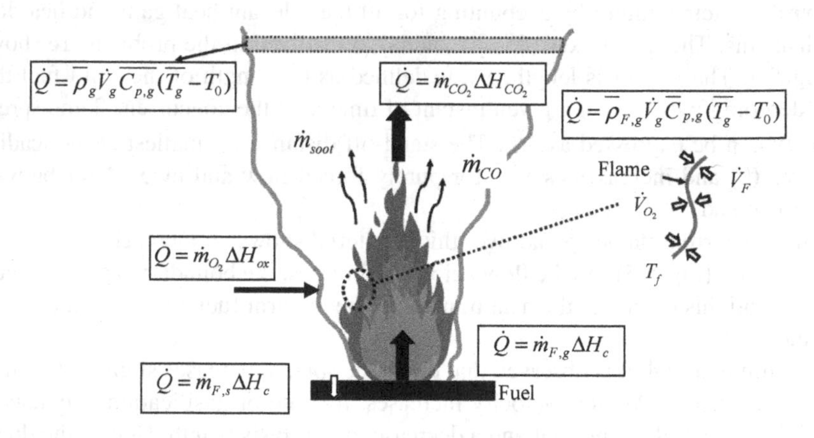

Fig. 5.6 The six different forms of burning rate expressions for fire

the burning rate can be expressed as $\dot{Q} = \dot{m}_{O_2} \Delta H_{ox}$, where $\Delta H_{ox} = 13.1$ kJ/g. This forms the basis of oxygen consumption calorimetry.

4. The energy release expressed in terms of kJ/g of CO_2 evolved. Similar to oxygen consumption, the burning rate can also be calculated if the rate of evolution of CO_2 is known. In this case, $\dot{Q} = \dot{m}_{CO_2} \Delta H_{CO_2}$. For most hydrocarbons, ΔH_{CO_2} is also a constant and equals 13.3 kJ/g. The accuracy can be improved by adding the rate of formation of soot and CO, thereby accounting for the efficiency in the combustion. In this case, $\dot{Q} = \dot{m}_{CO_2} \Delta H_{CO_2} + \dot{m}_{CO} \Delta H_{CO} + \dot{m}_{Carbon} \Delta H_{Carbon}$.
5. The energy released by combustion is used up to heat the gas to a flame temperature at the flame location. Thus, if radiation is ignored, the burning rate equals $\dot{Q} = \overline{\rho}_{F,g} \dot{V}_F \overline{C}_{P,g} (\overline{T}_f - T_0)$, where $\overline{\rho}_{F,g}$(kg/m^3) and $\overline{C}_{P,g}$ (kJ/kg-K) are the average density and specific heat of the gas near the flame, T_f (K) is the flame temperature, T_0 (K) is the ambient temperature, and $\dot{V}_F$ (m^3/s) is the volumetric flow rate of the fuel vapor reaching the flame. This equals the diffusion velocity of the fuel times the burning area. Since $\overline{\rho}_{F,g}$, $\overline{C}_{p,g}$, and T_f will not change much, the burning rate will scale linearly with the area of the flame, assuming that all the fuel is burned completely.
6. The energy released by calculating the temperature rise of the exhaust gases as shown in Fig. 5.6. If radiation is ignored, the burning rate equals

$$\dot{Q} = \dot{m}_a \overline{C}_{P,g} (\overline{T}_g - T_\infty) = \overline{\rho}_g \dot{V}_g \overline{C}_{P,g} (\overline{T}_g - T_\infty),$$

where $\dot{m}_a$ is the mass flow rate of gases entering the CV in the fire plume as shown in Fig. 5.6, $\overline{C}_{P,g}$ and $\overline{T}_g$ are the average specific heat and temperature of the gases. $\dot{m}_a$ can also be expressed as $\overline{\rho}_g \dot{V}_g$, where $\overline{\rho}_g$ is the average density of the exhaust gases (mostly air because of the high entrainment upstream) and $\dot{V}_g$ is the volumetric flow in m^3/s.

Out of the six methods, the first and second methods require knowledge of the mass loss rate (g/s) of the fuel vapor or the regression rate of the fuel surface in m/s times the density of the condensed fuel, and the heat of combustion ΔH_c. For many fuels, ΔH_c is listed in standard handbooks and ranges from 15 to 45 kJ/g. The mass loss rate changes with external factors such as wind and external heat flux such as the heat feedback from the walls and ceiling in a compartment fire. It will also change based on the oxygen mass fraction of the surrounding air. A reduction in the oxygen mass fraction will result in a lowering of the flame temperature and corresponding reduction in $\dot{Q}$. An in-depth mathematical analysis using the mass transfer number also called as the B-number to estimate the mass loss rate and heat release rate is discussed in the section below. Experimentally, the mass loss rate can be obtained by placing the fuel package on a weighing device and recording the mass as a function of time. The range of the mass loss rate per unit area typically varies from 2 g/m^{2}s to 100 g/m^{2}s. Below 2 g/m^{2}s usually leads to extinction.

Methods 3 and 4 can be used if the products of combustion can be collected in an exhaust duct and the concentration of O_2 and CO_2 is measurable. The heat release rate is then measured by knowing the mass flow rate of the exhaust gases multiplied by the oxygen mass fraction, and a constant equal to 13.1 kJ/g. The multiplication constant changes to 13.3 kJ/g if CO_2 is used instead of O_2. Methods 5 and 6 are based on the assumption that the energy released by combustion is used in raising the temperature of the gas in the control volume (the control volume around the flame in 5 and smoke plume in 6 as shown in Fig. 5.6). These methods are also called sensible enthalpy methods and were the first to be used in fire research to estimate the heat release rate. Method 6, for example, forms the basis of the Ohio State University Calorimeter [29]. It is also assumed that the fraction of heat that is used to evaporate or pyrolyze the fuel is negligible compared to the energy released. It should be noted that in methods 3–6, radiation from the flame is ignored. Radiation is usually added by a loss fraction denoted by χ_R. The resulting expression when used to calculate the heat release rate using method 6 is shown in Eq. (5.16).

$$\dot{Q} = \frac{\dot{m}_a \overline{C}_{p,g}}{1 - \chi_R} (\overline{T}_g - T_\infty). \tag{5.16}$$

The loss fraction, χ_R is usually calculated by doing a gas burner calibration. Values of χ_R varies between 0.1 and 0.5, as discussed in Chap. 3 (Eq. 3.39).

Heat Release Rate Theory For a flame to sustain over a fuel surface as shown in Fig. 5.7, the heat feedback from the flame should be sufficient to cause vaporization or pyrolysis of the liquid/solid fuel. The fuel vapor or pyrolyzate comes out at a

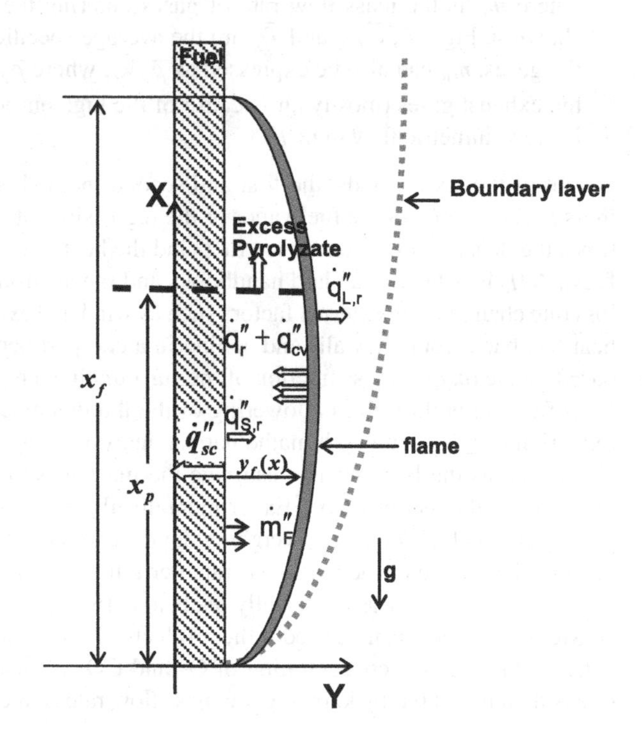

Fig. 5.7 Diffusion flame on a fuel surface

specific velocity known as the blowing velocity or Stefan mass transfer velocity. The fuel vapor then mixes with oxygen from the air to create a diffusion flame located at a distance $y_f(x)$ from the surface of the fuel as shown in Fig. 5.7. The burning rate per unit area ($\dot{m}''$) or mass burning flux is this velocity multiplied by the density of the fuel vapor ($\rho_{g,F} u_{F,g}$) or the density of the condensed fuel and the regression rate of the fuel surface ($\rho_{s,F} u_s$). Note that since the density of a liquid fuel or solid fuel is an order of magnitude higher than the corresponding gas vapor, $u_{F,g} >> u_s$.

The solution of the mass burning flux can be initiated by writing down a heat balance at the surface of the burning fuel

$$\dot{m}''L = \dot{q}'', \tag{5.17}$$

where L represents the latent heat of vaporization (for a liquid) or the heat of gasification (for a solid).

L can be expressed as

$$L = h_{fg} + C_{p,s}(T_s - T_0), \tag{5.18}$$

where h_{fg} is the heat of gasification (kJ/g), $C_{p,s}$ is the specific heat of the liquid or solid (kJ/g-K), T_s is the ignition temperature of the solid or the vaporization temperature of the liquid fuel and T_0 is the ambient temperature. $\dot{q}''$ is the net heat flux to the surface and depends on the nature of the flow field and boundary conditions (free stream temperature, ambient oxygen concentration *etc.*) It also depends on $\dot{m}''$, an effect called as the blocking effect [3]. If the net heat flux to the surface increases, then logically mass flux or mass loss rate per unit area will also increase. This causes the boundary layer to thicken causing reduction in gradients.

A larger burning rate will also place the flame farther from the surface thereby making the flame stand-off distance, $y_f(x)$ shown in Fig. 5.7, bigger. The flame is closer to the surface if the environment contains more oxygen or if the fuel is impure. The burning rate therefore does not increase linearly with external heat flux and to develop a general solution requires generalization of the heat flux to the fuel. It can be assumed that the heat flux to the surface is a product of a global heat transfer coefficient encompassing radiation (h_T) and temperature difference between the flame and the surface. If the flame temperature is denoted by T_f and surface temperature by T_s the mass loss rate per unit area can be represented as

$$\dot{m}'' = \frac{\dot{q}''}{L} = \frac{h(T_f - T_s)}{L} = B\frac{h}{C_{P,g}}, \tag{5.19}$$

where $B = \frac{C_{P,g}(T_f - T_s)}{L}$ is non-dimensional and represents the mass transfer number. h represents the convective heat transfer coefficient, T_f and T_s denote an average flame temperature and temperature of the fuel surface. The B-number (also called the Spalding mass transfer number) was first introduced by Spalding [23] in 1950 to characterize liquid fuel droplet burning and physically relates the heat release related to combustion (the numerator) to the losses associated with combustion (the denominator).

The heat transfer coefficient (h) is expressed in terms of a non-dimensional Nusselt number, defined as $Nu = \frac{hx}{k_g}$, where k_g denotes the thermal conductivity of the gas mixture at the interface between the condensed fuel and air and x is a characteristic length scale. In case of a flat fuel surface x can be defined as the distance from the leading edge and in the case of a sphere or cylinder refers to the diameter. The earlier expression (Eq. (5.19)) is now equal to:

$$\dot{m}'' = B\frac{k_g}{xC_P}Nu. \tag{5.20}$$

In order to account for the blocking effect, the Nusselt number is first evaluated without the blocking effect and then corrected for this effect through the ratio Nu/Nu_0

$$\dot{m}'' = B\frac{k_g}{xC_P}\frac{Nu}{Nu_0}Nu_0. \tag{5.21}$$

Non-dimensionalizing both sides by ρU_∞, where U_∞ is a characteristic velocity representing the flow field subjected to or induced by the diffusion flame, the above expression can be re-written as:

$$\frac{\dot{m}''}{\rho U_\infty} = B\frac{k_g}{C_{P,g}\mu_g}\frac{\mu_g}{\rho U_\infty x}Nu_0\frac{Nu}{Nu_0}. \tag{5.22}$$

$$\frac{\dot{m}''}{\rho U_\infty} = B(PrRe)^{-1}Nu_0\frac{Nu}{Nu_0}. \tag{5.23}$$

The quantity $\frac{k_g}{C_{p,g}\mu_g}$ equals 1/Pr where Pr = Prandtl number. The ratio $\frac{Nu}{Nu_0}$, the blocking effect can be calculated for several laminar flows and equals $\frac{ln(1+B)}{B}$ for small B. For turbulent flows it can be calculated empirically only. Typical values of B for many fuels are listed in Table 5.4. Expressions for Nu_0 for many geometrical configurations are available in standard heat and mass transfer textbooks and shown in Table 5.5.

For example, using Table 5.5, an expression for mass loss rate of a vertical plate in a laminar free convective flow can be calculated as:

$$\dot{m}'' = C_0B^m\frac{k_g}{C_{p,g}x}(GrPr)^{1/4}. \tag{5.24}$$

A mass burning rate for the same plate in a turbulent natural convective flow field is:

$$\dot{m}'' = C_0B^m\frac{k_g}{C_{p,g}x}(GrPr)^{1/3}. \tag{5.25}$$

Table 5.4 B-number values for different fuels [25]

Solids	Formula	B	Liquids	Formula	B
Polypropylene	C_3H_6	1.29	Methanol	CH_3OH	2.53
Polyethylene	C_2H_4	1.16	Ethanol	C_2H_5OH	2.89
Polystyrene	C_8H_8	1.55	Propanol	C_3H_7OH	3.29
Nylon 6/6	$C_{12}H_{22}N_2O_2$	1.27	Butanol	C_4H_8OH	3.35
Polycarbonate	$C_6H_{14}O_3$	1.41	*n*-pentane	C_5H_{12}	7.63
PMMA	$C_5H_8O_2$	1.78	*n*-hexane	C_6H_{14}	6.67
PVC	C_2H_3Cl	1.15	*n*-heptane	C_7H_{16}	5.92
Fir wood	$C_{4.8}H_8O_4$	1.75	*n*-octane	C_8H_{18}	5.42
α - cellulose	$C_6H_{10}O_5$	6.96	iso octane	C_8H_{18}	6.59
Polyoxymethylene	CH_2O	1.47	*n*-nonane	C_9H_2O	4.89
			n-decane	$C_{10}H_{22}$	4.61
			n-undecane	$C_{11}H_{24}$	4.43
			n-dodecane	$C_{12}H_{26}$	4.13
			Acetone	C_3H_2O	7.28

The constant C_0 can take different values and usually lies between 0.5 and 1.5. The value of the exponent m on B is ~0.5. Recently Ali et al. [24] have extended the correlation (laminar case) to include several orientation angles. The mass burning rate of a plate oriented at an angle θ with respect to the vertical for a plate in a free convective flow field is given by [24]:

$$\dot{m}'' = 0.737(Gr_x Pr_y)^{0.25} \frac{ln(1+B)}{B^{0.15}}, \tag{5.26}$$

where $Gr_x = \frac{g_{eff}\beta\Delta T x^3}{\nu^2}$ and $g_{\text{eff}} = g\cos\theta$ for $0 < \theta < 90°$ and $g_{\text{eff}} = g(\cos\theta - \sin\theta)$ for $-90° < \theta \le 0°$. Note that 0° represents the vertical case, −90° denotes a pool fire and 90° denotes a ceiling configuration. The mass loss rate per unit area can also be expressed as a regression rate $\dot{m}''/\rho_s$ where ρ_s is the density of the fuel (condensed phase). Typical regression rates of most fuels (solids/liquids shown in Table 5.4) vary between 0.02 and 0.4 mm/s. The regression rates are very small and are essentially due to the diffusion controlled nature of the problem. Table 5.5 shows the combination of Re or Gr, and Pr for flow configurations of interest in Fire that can be used to calculate the heat release rate.

The magnitude of the B-number reflects the differences in the fuel types. The B-number depends on the composition of the fuel vapor transferred as well as the free stream conditions. The independent fluid dynamic parameter is the Reynolds number for forced convection and the Grashof number for free convection. The Prandtl number for most flame gases can be assumed to be equal to the Schmidt number and is approximated to be equal to 1.

Table 5.5 Nu_0 values for some standard geometries and flow conditions: $Re = \frac{U_\infty x}{\nu_g}$ and $Gr = \frac{gx^3(T_f - T_0)}{T_0 \nu_g^2}$

Free convective flow		
Vertical plate (laminar, Gr < 10^9)	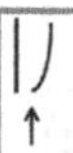	$0.59(Gr\,Pr)^{1/4}$
Vertical plate (turbulent, Gr > 10^9)	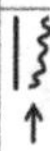	$0.14(Gr\,Pr)^{1/3}$
Horizontal plate burning face up (laminar, Gr < 10^7)	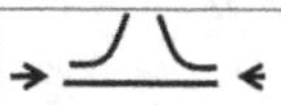	$0.54(Gr\,Pr)^{1/4}$
Horizontal plate burning face up (turbulent, Gr > 10^7)	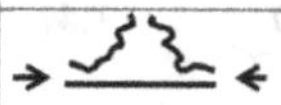	$0.14(Gr\,Pr)^{1/3}$
Horizontal plate burning face down (turbulent)		$0.27(Gr\,Pr)^{1/4}$
Horizontal cylinder		$0.525(Gr\,Pr)^{1/4}$
Sphere		$2 + 0.6Gr^{1/4}Pr^{1/3}$
Forced convective flow		
Horizontal flat plate (laminar)		$0.332Re^{1/2}Pr^{1/3}$
Horizontal flat plate (turbulent Re > 10^5)	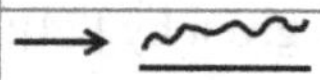	$0.036Re^{0.8}Pr^{0.3}$
Pool fire (laminar and axisymmetric)		$0.11Re^{1/2}Pr^{2/3}$
Droplet (laminar)		$0.37Re_d^{0.6}$
Cylinder (laminar)	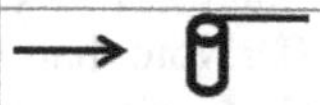	$0.891Re^{1/2}$
Cylinder (turbulent, Re > 40,000)		$0.27(Gr\,Pr)^{1/4}$

5.4 Flame Extinction

General strategies for extinguishing diffusion flames are cooling, reactant removal, chemical inhibition, and flame removal as shown in Fig. 5.8. The last three strategies attempt to approach flame extinction by influencing CV_1. The rate of gas-phase chemical reaction that was denoted by $\dot{\omega}_i'''$ plays an important role in flame extinction (as well as ignition).

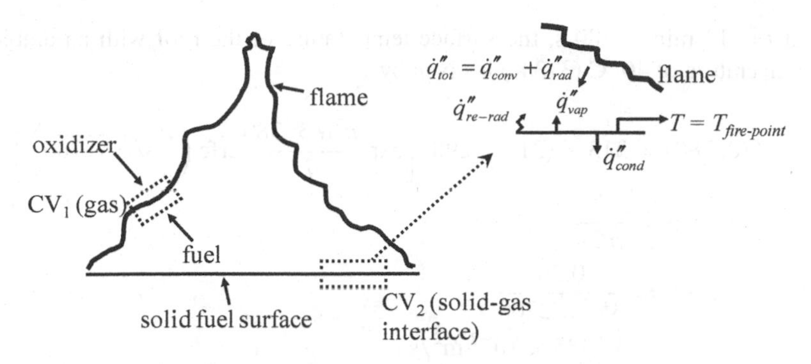

Fig. 5.8 Extinction of a diffusion flame by analyzing two control volumes. CV_1: around flame to determine limiting temperature, oxygen concentration criteria for flame extinction. And CV_2: around solid-gas interface determining limiting mass flux (critical mass flux) as a means for extinction

It is therefore necessary to formulate an expression to quantify $\dot{\omega}'''_i$ in order to establish critical extinction criteria. This is usually done by employing a one-step Arrhenius approximations to the rates. The analysis can be generalized using a non-dimensional Damköhler number [33], as discussed in Chap. 3.

5.5 Case Studies

1. A fire outbreak occurs in a forest which heats the ambient to a constant temperature of 40 °C, heating the wooden roof of a nearby home, initially at 25 °C, for 13 min through convective mode. Subsequently, a firebrand falls on its roof heating it constantly at 32.4 kW/m^2, with ambient air near the roof at 40 °C. Assuming that the convective heat transfer coefficient is 10 W/m^2-K, the approximate time available for evacuation from the home which is determined by the time to ignition of the roof surface can be evaluated. The properties of the roof material are given in the table below.

Ignition temperature	375 °C
Thermal conductivity	0.26 W/m-K
Density	354 kg/m^3
Specific heat	2200 J/kg-K
Emissivity	0.84

Stage I: Constant Convective Heating of the Roof by Ambient From Table 5.2, the surface temperature of the roof can be expressed as:

$$T(0,t) = T_\infty - (T_\infty - T_i)\left[\exp\frac{h^2\alpha t}{k^2}\text{erfc}\left(\frac{h}{k}\sqrt{\alpha t}\right)\right]$$

At $t = 13$ min $= 780$ s, the surface temperature of the roof with an ambient temperature of 40 °C (313 K) is given by:

$$T(0, 780) = 313 - (313 - 298)\left[\exp\frac{h^2\alpha \times 780}{k^2}\text{erfc}\left(\frac{h}{k}\sqrt{\alpha \times 780}\right)\right]$$

$$\begin{aligned}\alpha &= \frac{k}{\rho C_P}\\ &= \frac{0.26}{354 \times 2200}\\ &= 3.3385 \times 10^{-7}\ \text{m}^2/\text{s}\\ \frac{h^2\alpha}{k^2} &= 4.9386 \times 10^{-4}\ s^{-1}\\ \frac{h\sqrt{\alpha}}{k} &= 0.0222\, s^{-1/2}\\ T(0, 780) &= 313 - (313 - 298)\left[\exp(0.3852)\text{erfc}(0.62)\right].\end{aligned}$$

Solving the above equation, we have $T(0, 780) = 304.6$ K. Thus, the roof will slowly reach a temperature of 304.6 K or 31.6 °C because of convective heating by the ambient for a duration of 13 min. After this, localized heating will occur because of the fire brand. The rise in temperature is solved next.

Stage II: Constant Surface Heat Flux and Convective Heat Transfer From Table 5.2, the time for the surface of the roofing material to reach an ignition temperature of 375 °C (648 K) can be obtained by solving the equation below:

$$\begin{aligned}T_{ig} &= T(0, t_{ig})\\ &= T_i + \frac{\dot{q}_s''}{h_T}\left[1 - \exp\left(\frac{h_T^2\alpha t_{ig}}{k^2}\right)\text{erfc}\left(\frac{h_T\sqrt{\alpha t_{ig}}}{k}\right)\right].\end{aligned}$$

Observe that the heat flux from the fire brand is included in the above equation as $\dot{q}_s''$.

The overall heat transfer coefficient is given by:

$$\begin{aligned}h_T &= h + \frac{\epsilon\sigma\left[T(0, t_{ig})^4 - T_\infty^4\right]}{T(0, t_{ig}) - T_\infty}\\ &= 10 + \frac{0.84 \times 5.67 \times 10^{-8}(648^4 - 313^4)}{648 - 313}\\ &= 33.7033\ \text{W/m}^2\text{-K}\end{aligned}$$

$$648 = 304.6 + \frac{32400}{33.7033}\left[1 - \exp(5.6098 \times 10^{-3}t_{ig})\text{erfc}\left(0.0749\sqrt{t_{ig}}\right)\right]$$

$$f(t_{ig}) = \left[1 - \exp(5.6098 \times 10^{-3}t_{ig})\text{erfc}\left(0.0749\sqrt{t_{ig}}\right)\right] - 0.643 = 0.$$

t_{ig} (min)	t_{ig}(s)	f(t_{ig})
1	60	0.066
3	180	0.0692
5	300	−0.0011
6	360	−0.022

The table above shows that in approximately 5 min (300 s), the temperature of the roofing material will equal the ignition temperature after the fire brand falls on the roof. The roof will thus ignite in 300 s.

2. A baking pan is placed on a metal tray in a conventional baking oven. While placing the baking pan, a finger touches the tray which is at 120 °C. Assuming a human reaction time of 1.05 s to remove the finger, the depth up to which a temperature rise will be seen can be evaluated. Additionally, an ice pack is applied on the burn area removing heat at a rate of 6 kW/m^2. The time required to restore the skin layer to normal temperature of 37 °C can be determined. The thermal properties are given in the table below.

Property	Skin	Metal tray
Thermal conductivity (W/m-K)	0.37	200
Density (kg/m^3)	1000	2700
Specific heat (J/kg-K)	3770	860

Stage I: Contact between Two Solids [Skin (A) and Metal Tray (B)] From Table 5.2, the temperature at the contact point between the two solids can be determined as:

$$T_{i,0} = T(0,0) = \frac{\sqrt{k_A \rho_A C_A} T_{i,A} + \sqrt{k_B \rho_B C_B} T_{i,B}}{\sqrt{k_A \rho_A C_A} + \sqrt{k_B \rho_B C_B}}.$$

Using the properties for skin and metal tray, the interface temperature is given as:

$$T_{i,0} = 388.69\,\text{K}$$

Now, we need to see the depth upto which the effect of contact between skin and metal tray is observed within solid A (skin). This can be done by evaluating x, where $T = 37\,°\text{C}$ to determine depth upto which temperature rise is seen. Using the equation in Table 5.2 for temperature variation in solid A:

$$\frac{310 - 310}{388.69 - 310} = 1 - \text{erf}\left(\frac{x_A}{2\sqrt{9.8143 \times 10^{-8} \times 1.05}}\right) \Rightarrow x_A = 1.284\,\text{mm}.$$

This shows that the effect of the hot metal tray is felt upto a depth of 1.284 mm. **Stage II: Constant Surface Heat Flux** Additionally, an ice pack is used which is considered as a constant heat flux sink. Using Table 5.2, the time required for the skin surface to attain the normal skin temperature (37 °C) can be evaluated.

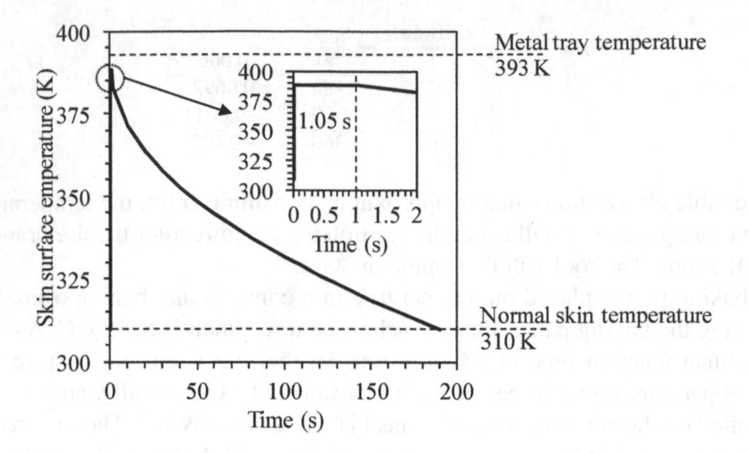

Fig. 5.9 Time evolution of temperature

$$T(x,t) - T_i = \frac{\dot{q}_s''}{k}\left[2\sqrt{\frac{\alpha t}{\pi}}\exp\left(-\frac{x^2}{4\alpha t}\right) - x\text{erfc}\left(\frac{x}{2\sqrt{\alpha t}}\right)\right].$$

At skin surface where $x = 0$,

$$310 - 388.69 = \frac{-6\times 10^3}{0.37}\left[2\sqrt{\frac{9.8143\times 10^{-8}t}{\pi}}\right] \Rightarrow t = 188.4\,\text{s}$$

This shows that it takes 188.4 s (≈3.14 min) to cool the skin surface to 37 °C (310 K).

The time evolution of temperature for the skin surface is shown in Fig. 5.9. Initially there is a sudden increase in temperature due to surface contact between the skin and the tray at 120 °C. After 1.05 s, the ice pack is applied. This results in gradual decrease in skin surface temperature to 37 °C after about 3 min.

3. A corrugated cardboard sheet initially at a temperature of 300 K is subjected to constant surface heat flux of 12 kW/m^2. Its time to ignition can be determined as the ignition temperature is known to be 350 °C. The properties of the cardboard are given below.

Property	Value
Porosity	0.6
Thermal conductivity	0.064 W/m-K
Density	300 kg/m^3
Specific heat	1520 J/kg-K
Activation energy	1.3×10^5 J/mol
Pre-exponential factor	10^{19} s^{-1}

Corrugated cardboard is a porous reacting solid and the time to ignition is dependent on its thermal properties, porosity, as well as its reaction parameters. Thus, using the equation from Table 5.2 by Telengator et al. [16],

$$t_{ig} = \frac{\pi k \rho C T_i^2}{4\dot{q}_e''} \left[\frac{\frac{E}{RT_i}}{\ln \frac{k_s \rho_s C_s T_{ig}^2 A}{0.65(\dot{q}_e'')^2 \left(\frac{E}{RT_i}\right)^{1/2}} \left[\frac{\pi(\dot{q}_e'')^2 t_{ig}}{k_s \rho_s C_s T_i^2}\right]^{1/4}} - 1 \right]^2 (1 - \epsilon_P)$$

$$\times \left(1 - \epsilon_P + \frac{\lambda_g \epsilon_P}{\lambda_s}\right).$$

The time to ignition can be evaluated as shown below:

$$t_{ig} = 171907.95 \left[\frac{52.12}{\ln\left(1.0798 \times 10^{20} (t_{ig}^{1/4})\right)} - 1 \right]^2 \times 0.2575.$$

It is seen that a non-linear equation needs to be solved to evaluate the time to ignition. This can be solved iteratively by equating the following equation to zero.

$$f(t_{ig}) = t_{ig} - 171907.95 \left[\frac{52.12}{\ln\left(1.0798 \times 10^{20} (t_{ig}^{1/4})\right)} - 1 \right]^2 \times 0.2575 = 0.$$

Solving iteratively,

t_{ig} (s)	$f(t_{ig})$
500	116.55
400	5.925
395	0.321

Time to ignition = 395 s. It is seen that it takes 395 s to ignite the corrugated cardboard when a constant heat flux of 12 kW/m^2 is applied.

4. Due to COVID, a machine shop installs plastic barriers to keep employees safe during business hours, as shown in Fig. 5.10. The safety engineer decides to perform a fire risk analysis imposed by the barrier and wants to know at what rate the fire spread given a condition where the top corner of the plastic barrier ignites and a flame propagates downward and laterally on either side as shown in Fig. 5.10.

   (a) An expression for the heat release rate as a function of velocity, heat of combustion, and mass burning rate per unit area can be developed.
   (b) The velocity of flame propagation and the plot of the heat release rate versus time can be determined.

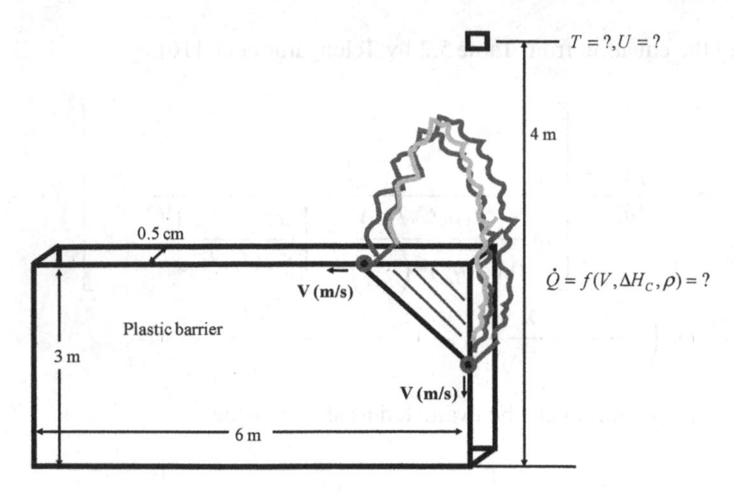

Fig. 5.10 Schematic of flame spreading on a thermally thin plastic sheet ignited at the top corner

(c) The response time of the heat detector located exactly above the fire and 4 m from the floor can be evaluated given that the detector activation temperature equal to 120 °C, and the Response Time Index (RTI) is 150 (m-s)$^{1/2}$.

Properties are listed below:

$\Delta H_c = 30$ kJ/g,
$L = 1.65$ kJ/g,
$C_s = 0.8$ kJ/kg-K,
$k_s = 0.26$ W/mK,
$\rho_s = 860$ kg/m^3,
$T_P = 350\,°$C,
$T_\infty = 20\,°$C,
$T_f = 2000\,°$C.

Air: $k_g = 0.03$ W/m-K, $\nu_g = 1.54 \times 10^{-5}$ m^2/s. For all calculations, the plastic barrier is assumed to be thermally thin. B number is 2.1 and the flame temperature is assumed to be 2000 °C. For the vertical configuration, the convective heat transfer coefficient is assumed to be a constant of 20 W/m^{2}K. Losses due to radiation are neglected.

For thermally thin case, the flame spread velocity is not a function of u_∞ and from Table 5.3:

$$V_F = \frac{\sqrt{2}k_g(T_f - T_{vap})}{\rho_s C_s \delta(T_{vap} - T_\infty)}$$

$$V_F = \frac{\sqrt{2} \times 0.03 \times (20000 - 350)}{860 \times 0.8 \times 10^3 \times 0.25 \times 10^{-2} \times (350 - 20)}$$
$$= 1.23 \times 10^{-4}$$
$$= 0.123 \text{ mm/s}$$

Note that although thickness of the barrier is 0.5 cm as shown in Fig. 5.10, the thickness δ used to estimate V_F is half (0.25 cm) because the flame propagates on both sides of the barrier. If the barrier thickness is reduced to half ($\delta = 0.25/2 = 0.125$ cm), the flame velocity doubles to 0.246 mm/s. Thus the fire hazard increases when the thickness is reduced. Note that the flame spread velocity is a low value. If the bottom corner of the plastic were ignited and flame spread vertically upwards, then velocity would have been 10–100 times faster! The area of the pyrolysis/burning region with both sides burning is shown in the sketch below:

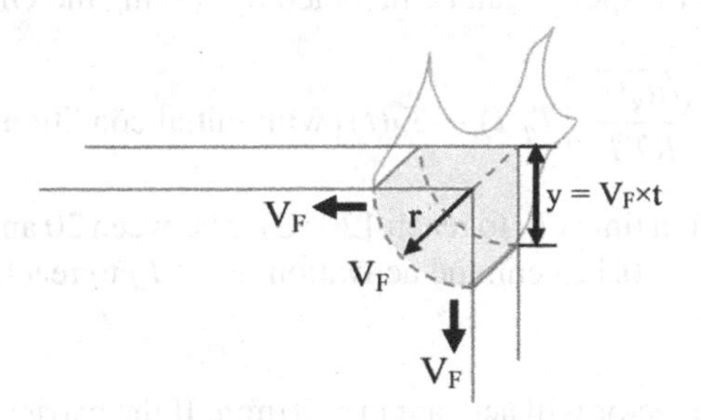

The burning area is a quarter of a circle with radius of $y = V_F \times t$:

$$A(t) = 2 \times \frac{1}{4} \times (\pi r^2)$$
$$= \frac{\pi}{2} V_F^2 t^2$$

Neglecting the thickness of the slab, this area will be burning with a mass loss rate per unit area, $\dot{m}''_F$:

$$\dot{m}''_F = \frac{h}{C_P} \ln(1 + B)$$

$$\dot{Q} = \dot{m}''_F A(t) \Delta H_c$$

$$\dot{Q}(t) = \frac{\pi}{2} \frac{h V_F^2}{C_P} \ln(1 + B) t^2 \Delta H_c$$

where h, C_P, B are given constants, and V_F is the velocity of the flame propagation. Thus $\dot{Q}$ will follow a "square" fire growth rate such that $\dot{Q} = \alpha t^2$ where

α is a constant and equal to

$$\frac{\pi}{2}\frac{hV_F^2}{C_P}\ln(1+B)\Delta H_c$$

The heat release rate, $\dot{Q}$ can be estimated with time iteratively.

The detector is located exactly $z = 1$ m above the fire. We can use the Zukoski plume correlation (Chap. 6) to estimate T_g and u_g at the detector:

$$T_g(t) = T_\infty + 13.74\left[\frac{\dot{Q}(t)^{2/5}}{z}\right]^{5/3}$$

$$u_g(t) = 0.59\left[\frac{\dot{Q}(t)}{z}\right]^{1/3}.$$

The detector response can be modeled by solving the ODE:

$$\frac{dT_d}{dt} = \frac{\sqrt{u_g(t)}}{RTI}\left[T_g(t) - T_d(t)\right] \text{ with initial condition: } t = 0,\ T_d(t) = T_\infty$$

The activation time (T_d to reach 120 °C) is between 20 and 21 min. For thickness $\delta = 0.25/2 = 0.125$ cm: the activation time (T_d to reach 120 °C) is between 11 and 12 min.

Thus, the detector will activate in ~20 min. If the barrier thickness is reduced to 0.25 cm then $\delta = 0.125$ and the detector will now activate in 11 min. This case study is also useful in showing how the flame spread rate differs from the burning rate. Both the downward spread on the thin plastic sheet and the corresponding burning rate are evaluated. It should be clear to the student that the spread rate is the local velocity at which the boundary of the burning area moves over the unburned fuel. The burning rate is (the velocity of gas vapor undergoing combustion × the gas phase density of the fuel vapor) or (the velocity at which the plastic surface regresses × the density of the plastic) multiplied by the heat of combustion.

Review Questions

1. Considering the ignition of trees in a forest fire, explain how the following factors could affect the time to ignition: (a) wind speed, (b) rainfall, (c) strength of lightning activity, and (d) forest cover density.
2. For a semi-infinite PMMA solid with incident heat flux of 10 kW/m^2 on its side, determine the time to ignition. The slab is initially at 298 K. The thermal

diffusivity and thermal inertia of PMMA are given as 1.01×10^{-7} m^2/s and 0.73 (kW/m^{2}K)^2-s, respectively. The ignition temperature is 278 °C. Find the ambient temperature such that the convective heating ignites it in the same time. Take heat transfer coefficient to be 25 W/m^2-K.

3. A heated copper rod is in contact with Douglas fir wood initially at 20 °C. Determine the minimum temperature (T_{min}) for copper rod such that the wood surface will ignite. Also, determine the length upto which the effect of copper rod will be felt within the wood after 10 min when copper rod surface is at temperature of T_{min}. The values of thermal inertia of copper and fir wood are 1325 (kW/m^{2}K)^2-s and 0.25 (kW/m^{2}K)^2-s, respectively. The ignition temperature and thermal diffusivity of the wood can be taken as 384 °C and 0.08 mm^2/s,, respectively.
4. Consider a solid block of 1 mm thickness with thermal conductivity of 0.4 W/m-K, initially at 20 °C. The block is exposed to ambient air at 600 K on both sides. The convective heat transfer coefficient is 10 W/m^2-K and the block is exposed for 6 min. Determine if the solid would ignite if its ignition temperature is 200 °C and thermal diffusivity is 10^{-7} m^2/s.
5. What is the difference between opposed and concurrent flame spread.
6. What is the significance of flame heating length and how is it different from pyrolysis length?
7. Plot the variation in the flame spread rate for a thermally thick solid as a function of (a) the pyrolysis temperature, (b) air flow velocity in forced convection, (c) heat of combustion of fuel for flame spread in natural convection, and (d) thermal inertia of the solid.
8. What are the different ways of evaluating the burning rate? Also, determine mass loss rate of the fuel during PMMA combustion if 11 g/s of CO_2 is measured in the exhaust.
9. What is the physical significance of B number in the definition of mass loss rate? What conclusion can be drawn regarding the mass burning rate given that cellulose has a much higher B number as compared to PMMA?
10. A vertically placed PVC pipe (diameter = 1 cm, length = 10 cm) weighs 11 g initially. It burns in a garden where wind blows at a speed of 0.4 m/s. Evaluate the time taken for it to burn completely. The gas properties can be evaluated at an average temperature of 1300 K.

References

1. Torero, J. 2016. Flaming ignition of solid fuels, *SFPE handbook of fire protection engineering*, 633–661. Springer.
2. Annamalai, K., and P. Durbetaki. 1976. Ignition of thermally thin porous pyrolyzing solids under normally impinging flames. *Combustion and Flame* 27: 253–266.
3. Quintiere, J.G. 2006. *Fundamentals of Fire Phenomena*. New York: Wiley.
4. Babrauskas, V. 2003. *Ignition Handbook*. Issaquah, WA: Fire Science Publishers.

5. Raghavan, V., A. Rangwala, and J. Torero. 2009. Laminar flame propagation on a horizontal fuel surface: Verification of classical Emmons solution. *Combustion Theory and Modelling* 13 (1): 121–141.
6. Quintiere, J.G. 1981. A simplified theory for generalizing results from a radiant panel rate of flame spread apparatus. *Fire and Materials* 5 (2): 52–60.
7. Drysdale, D. 2011. *An Introduction to Fire Dynamics*. Wiley.
8. Atreya, A., and I. Wichman. 1989. Heat and mass transfer during piloted ignition of cellulosic solids. *Journal of Heat Transfer* 111: 719–725.
9. Roberts, A.F., and B.W. Quince. 1973. A limiting condition for the burning of flammable liquids. *Combustion and Flame* 20 (2): 245–251.
10. Rasbash, D.J., and D.D. Drysdale. 1983. Theory of fire and fire processes. *Fire and Materials* 7 (2): 79–88.
11. Quintiere, J., and A. Rangwala. 2004. A theory for flame extinction based on flame temperature. *Fire and Materials* 28 (5): 387–402.
12. Atreya, A. 1998. Ignition of Fires. *Philosophical Transactions of the Royal Society of London A* 356: 2787–2813.
13. Nellis, G., and S.A. Klein. 2020. *Introduction to Engineering Heat Transfer*. Cambridge University Press.
14. Quintiere, J.G. 1981. A simplified theory for generalizing results from a radiant panel rate of flame spread apparatus. *Fire and Materials* 5 (2): 52–60.
15. Liñán, A., and F. Williams. 1971. Theory of ignition of a reactive solid by constant energy flux. *Combustion Science and Technology* 3 (2): 91–98.
16. Telengator, A.M., S.B. Margolis, and F.A. Williams. 1999. Analysis of ignition of a porous energetic material. *Combustion Theory and Modelling* 3 (1): 33–49.
17. Vilyunov, V.N., V.E. Zarko, O. Zhitnitskaya, and V. Bukharov. 1989. Ignition of Solids. *Studies in Physical and Theoretical Chemistry*, 60.
18. Beever, P.F. 1995. Self-heating and spontaneous combustion. *The SFPE Handbook of Fire Protection Engineering* 2: 2–189.
19. Fernandez-Pello, A.C. 1995. The Solid Phase. In *Combustion Fundamentals of Fire*, ed. G. Cox, 32–100. New York: Academic Press.
20. de Ris, J. 1969. The Spread of a Laminar Diffusion Flame. *Proceedings of the Combustion Institute* 12: 241–252.
21. Altenkirch, R., R. Eichhorn, and P. Shang. 1980. Buoyancy effects on flames spreading down thermally thin fuels. *Combustion and Flame* 37: 71–83.
22. Babrauskas, V., and R.D. Peacock. 1992. Heat release rate: the single most important variable in fire hazard. *Fire Safety Journal* 18 (3): 255–272.
23. Spalding, D.B. 1950. Combustion of liquid fuel in gas stream. *Fuel* 29: 2–7.
24. Ali, S.M., V. Raghavan, and A.S. Rangwala. 2010. A numerical study of quasi-steady burning characteristics of a condensed fuel: effect of angular orientation of fuel surface. *Combustion Theory and Modelling* 14 (4): 495–518.
25. Annamalai, K., and M. Sibulkin. 1979. Flame spread over combustible surfaces for laminar flow systems part I: excess fuel and heat flux. *Combustion Science and Technology* 19 (5): 167–183.

Chapter 6
Properties of Fire Plumes

Flow fields in scenarios starting from simple gaseous jet diffusion flames to fires established over condensed (liquid and solid) fuels are dominated by buoyancy-driven flow of various degrees. In the case of gaseous fuel, the gas jet issued from a burner has its initial momentum, which is generally much higher than the momentum of the fuel stream generated over the condensed fuel surfaces. Based upon the initial momentum of the fuel stream, the buoyancy effects set in. The buoyancy force is induced due to density differences (caused by temperature gradients) in the flow field. As a result, the hotter gases rise upwards due to their relative lower density (caused by higher temperature) when compared to that of the atmospheric air. This buoyancy-driven flow of hot gases established over a fire is often termed as a fire plume. It starts as a laminar flow and subsequently transitions to the turbulent regime, depending upon the strength and dimension of the fire. The buoyancy force decreases as the density differences in the flow field decreases. This happens at a certain height above the fire. The temperature and velocity variations within this height characterize a fire plume. As the hot gases rise upwards, cold atmospheric air and other gases entrain into the fire plume, due to the fluid friction (viscosity) in laminar flows and predominantly due to the action of the eddies in turbulent flows. Eddies of different scales are formed over the cold edges of the fire plume and they affect the entrainment process significantly. This entrainment rate is another characteristic parameter of a fire plume. Due to eddy dynamics the fire plume is seen to be oscillatory.

6.1 Characteristics of Fire Plumes

In this section, the characteristics of a buoyant fire plume are revealed starting from a basic procedure outlined in Quintiere [1] and Heskestad [2]. A fire plume has three zones as observed by several researchers (Fig. 6.1).

A. S. Rangwala and V. Raghavan, *Mechanism of Fires*,
https://doi.org/10.1007/978-3-030-75498-3_6

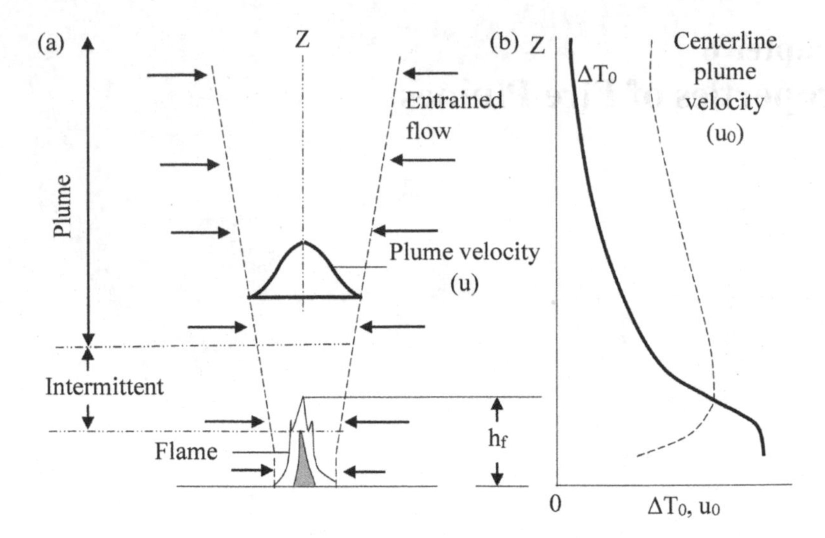

Fig. 6.1 **a** Schematic of different zones of a fire plume (flame, intermittent, and plume), and **b** variation of excess temperature (ΔT_0) and centerline plume velocity (u_0) with height (Z)

The zone where the heat release occurs is called the flame zone and this prevails near the fuel surface. This zone extends to an average flame height. Above this zone, an intermittent zone prevails. Here, the flame is seen intermittently. This zone encompasses the maximum flame length, which is almost twice of the minimum flame height. Significant fluctuations are observed in this zone. Over this zone only hot gas prevails and is called the plume zone, as shown in Fig. 6.1. The flame height is determined by plotting a variable called intermittency, I, which is the ratio of the time for which a flame is present in a given location over the total time of observation. In the flame zone, the value of I is unity. In the intermittent zone, the value of I reduces sharply and reaches a value of zero at the plume zone. The mean flame length is often fixed as the axial location, at which the value of I is 0.5.

The net energy content in the gases increases along the height from the fuel surface due to heat release from the flame and becomes almost a constant around the tip of the flame (at the average flame height). The energy content of the plume gases is the heat released due to combustion minus the heat lost to the environment. The heat loss is often estimated as a fraction of the heat release rate. Similarly, the average temperature increases with height, from the fuel surface, and reaches the maximum value equal to the flame temperature. This value remains almost a constant in the flame zone. The temperature decreases in the intermittent and the plume zones.

The characteristic non-dimensional number that characterizes the plume behavior is called the Froude number. As discussed in Chap. 1 it comes from the non dimensional momentum conservation, and expressed as $\mathrm{Fr} = U^2/(gD)$, where U is the characteristic velocity, taken as the average plume velocity, D is the characteristic length scale, usually taken as the diameter of the plume source, and g is the acceleration due to gravity. An estimate of the buoyant flow velocity at an elevation of z from the fuel surface is obtained by comparing the buoyancy (relative density) term $(\rho_\infty - \rho)gz$ to the gain in the kinetic energy $0.5\rho U^2$. Here, the subscript ∞ denotes the ambient condition. For open plumes, the pressure variation along the plume is quite negligible. Thus, the density is written as an inverse of temperature and U is estimated as

$$U \approx [2(T - T_\infty)gz/T_\infty]^{0.5},$$

where T is the average radial temperature of the plume at height z.

For fires over condensed fuels, the vaporized or gasified fuel coming out of the fuel surface has a velocity of the order of few cm/s that is significantly smaller than the plume velocity established by the buoyancy forces. The entrainment of air is caused by much higher plume velocity. It should be noted that in the case of a jet diffusion flame of a gaseous fuel, the jet velocity dictates the entrainment rate. As the plume temperature decreases, the average velocity of the plume also decreases. Thus, the rate of air entrainment decreases with plume height. The air entrained up to the flame height is used for burning the fuel.

Since in the flame zone, the average temperature remains constant, the rate of mass of air entrained is approximately estimated in terms of pool/plume circumference, height of the flame (h_f), and average entraining air velocity (U_e) as,

$$\dot{m}_a \approx \rho_\infty \times \pi \times D \times h_f \times U_e.$$

Since the entrainment rate is dependent on the plume velocity (U), it is noted from the estimation of U that the maximum plume velocity is proportional to $[g \times z]^{0.5}$. Using the flame height, it is noted that the plume velocity and, as a result, the entrainment velocity is proportional to $[g \times h_f]^{0.5}$. Diffusion flames form when the fuel and oxidizer mix in stoichiometric proportion. If s is the stoichiometric air-to-fuel ratio, then

$$s = \dot{m}_a/\dot{m}_F.$$

Here, $\dot{m}_F$ is the mass burning rate of the fuel. Using the estimations of $\dot{m}_a$ and U_e:

$$s \approx \frac{\rho_\infty \times \pi \times D \times h_f \times U_e}{\dot{m}_F},$$

$$s \approx \frac{\rho_\infty \times \pi \times D \times h_f \times \sqrt{gh_f}}{\dot{m}_F},$$

or

$$\frac{h_f}{D} \approx \left[\frac{\dot{m}_F \times s}{\pi \rho_\infty D^2 \sqrt{gD}}\right]^{2/3}.$$

For a large fire of diameter 1 m, $(T - T_\infty)/T_\infty$ 10, g 10 m/s2 @ 1/2 m from the fire, velocity 10 m/s using above equation. The heat release rate ($\dot{Q}$) is given by $\dot{Q} = \dot{m}_F \times \Delta H_c$, where ΔH_c is the heat of combustion. Using this, the expression for non-dimensional flame height can be written as,

$$\frac{h_f}{D} \approx \left[\frac{s c_{p\infty} T_\infty}{\pi \Delta H_c}\right]^{2/3} \left[\frac{\dot{Q}}{\rho_\infty c_{p\infty} T_\infty D^2 \sqrt{gD}}\right]^{2/3} \approx C(Q^*)^{2/3}.$$

Here, Q^* is the non-dimensional quantity called Zukoski number. The non-dimensional constant C is defined as $\left[\frac{s c_{p\infty} T_\infty}{\pi \Delta H_c}\right]^{2/3}$. The denominator in the definition of Q^* represents the thermal energy transported by the plume. Momentum-dominated jet flame regimes are represented by value of Q^* greater than 10^4, as indicated by McCaffrey [3]. Q^* represents ratio of combustion energy to nominal plume energy. It can also be used to represent a ratio of a length scale as shown in Eq. (6.1).

$$(Q^*)^{2/5} = \left[\frac{\dot{Q}}{\rho_\infty c_{p\infty} T_\infty D^2 \sqrt{gD}}\right]^{2/5} = \left[\frac{\dot{Q}}{\rho_\infty c_{p\infty} T_\infty \sqrt{g}}\right]^{2/5} / D. \tag{6.1}$$

Therefore, Heskestad [2, 4] reported the correlation in Eq. (6.2) that works for a wide range of naturally occurring fires given as,

$$h_f = 0.23(\dot{Q})^{2/5} - 1.02D, \tag{6.2}$$

where h_f and D are in m, and $\dot{Q}$ is in kW.

6.2 Theoretical Analysis of Idealized Fire Plumes

An idealized plume has a point source from which a buoyant flow generates as shown in Fig. 6.2 (adapted from Karlsson and Quintiere [5]). Simplifying assumptions given below are invoked to obtain a theoretical framework for the problem.

(1) The change in the density within the plume region is negligible when compared to the ambient density. This allows the usage of ambient density for the plume density. However, the buoyancy force is calculated using the difference between the ambient and the plume density. This is also known as Boussinesq approximation.
(2) No heat losses due to radiation.
(3) The profiles of velocity and temperature are uniform and a top-hat profile is assumed.
(4) The velocity of air entraining through the boundary of the plume is proportional to the plume velocity at the given height (consider $v = \alpha u$), as shown in Fig. 6.2.
(5) Symmetric domain.

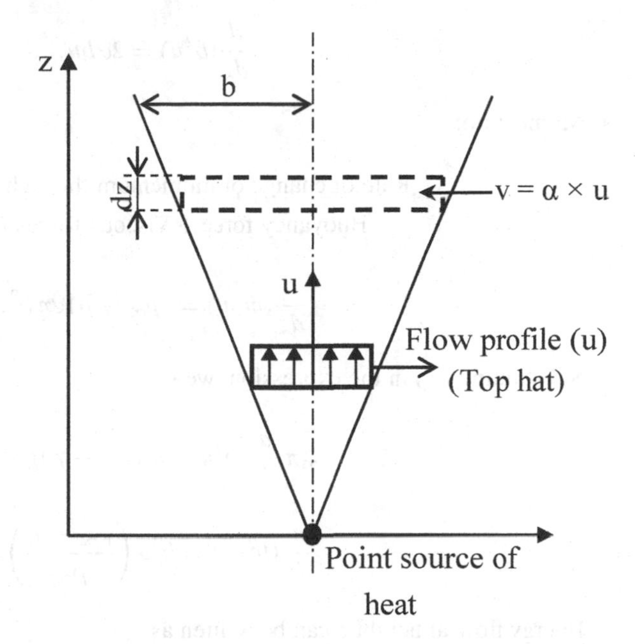

Fig. 6.2 Schematic of an idealized fire plume [5]

(6) No viscous effects.

Following Morton et al. [6], the mean flow variables in an idealized plume are calculated by solving the conservation equations for mass, momentum, and buoyancy, which are given as

- Mass: The mass flow rate (kg/s) in a differential element shown by dz is equal to

$$\dot{m}_p = \rho u \pi b^2 \quad \text{or} \quad \dot{m}_p(z) = \rho u(z)\pi [b(z)]^2,$$

where b is the radius of plume at height z and u is the vertical velocity at z as shown in Fig. 6.2. The conservation of mass can now be written as

Increase in mass flowing up through differential element dz
= Mass entrained through sides of dz per unit height.

$$\frac{d}{dz}(\dot{m}_p) = \frac{2\pi b(dz) v \rho_\infty}{dz}.$$

Here, $2\pi b(dz)$ is the circumferential area through which air entrains into the differential element dz. Substituting $\dot{m}_p$ in the expression:

$$\frac{d}{dz}(\rho_\infty u \pi b^2) = 2\pi b \alpha u \rho_\infty.$$

On simplification,

$$\frac{d}{dz}(b^2u) = 2\alpha bu. \tag{6.3}$$

- Momentum:

Rate of change of momentum through height dz
= Buoyancy force + Viscous forces (~ 0)

$$\frac{d}{dz}(\dot{m}_p u) = (\rho_\infty - \rho)g\pi b^2.$$

Substituting $\dot{m}_p$ in the expression, we get

$$\rho_\infty \pi \frac{d}{dz}(b^2u^2) = (\rho_\infty - \rho)g\pi b^2.$$

$$\frac{d}{dz}(b^2u^2) = b^2 g\left(\frac{\rho_\infty - \rho}{\rho_\infty}\right). \tag{6.4}$$

Energy flow at height z can be written as

$$\dot{Q}_c = \dot{m}_p C_p \Delta T = \pi b^2 \rho u C_p \Delta T, \tag{6.5}$$

where $\Delta T = T(z) - T_\infty$, and T is the temperature at height z. Since a top-hat profile is assumed, T does not vary along the plume radius and only varies with z.
- From Ideal gas law,

$$\rho T = \rho_\infty T_\infty.$$

Using $\Delta T = T - T_\infty$ and $\Delta\rho = \rho_\infty - \rho$,

$$\frac{\Delta\rho}{\rho_\infty} = \frac{\Delta T}{T} = \left(\frac{T - T_\infty}{T_\infty}\right)\left(\frac{T_\infty}{T}\right).$$

$$\frac{\Delta\rho}{\rho_\infty} \approx \left(\frac{T - T_\infty}{T_\infty}\right) \text{ when } T/T_\infty \to 1. \tag{6.6}$$

In Eqs. (6.3)–(6.6), g is the acceleration due to gravity, ρ is the plume density, and ρ_∞ is the ambient density. Using Eq. (6.6) in Eq. (6.4):

$$\frac{d}{dz}(b^2u^2) = b^2 g\left(\frac{\Delta T}{T_\infty}\right). \tag{6.7}$$

Writing ΔT in terms of $\dot{Q}$ from Eq. (6.5):

$$\frac{d}{dz}(b^2u^2) = \left(\frac{\dot{Q}_c g}{\pi \rho_\infty u C_p T_\infty}\right). \tag{6.8}$$

Equations (6.3) and (6.8) form two ordinary differential equations with unknowns b and u. Given a $\dot{Q}_c$, they can be solved simultaneously to give the value of plume radius and velocity at any height z. The entrainment velocity at any height is equal to αu, where α is a constant based on experiments. For solving the ordinary differential Eqs. (6.3) and (6.8), assume $b = C_1 z^m$ and $u = C_2 z^n$. From Eq. (6.3):

$$\frac{d}{dz}(C_1^2 z^{2m} C_2 z^n) = 2C_1 z^m \alpha C_2 z^n.$$

$$C_1^2 C_2 (2m + n) z^{2m+n-1} = 2C_1 C_2 \alpha z^{m+n}.$$

From Eq. (6.8):

$$C_1^2 C_2^2 (2m + 2n) z^{2m+2n-1} = \left(\frac{\dot{Q}_c g}{\pi \rho_\infty C_2 z^n C_p T_\infty}\right).$$

Equating powers, we get,

$$m = 1 \text{ and } n = -1/3$$

$$C_1 = \frac{6}{5}\alpha \quad \text{and} \quad C_2 = \frac{25}{48\alpha^2} \frac{\dot{Q}_c g}{\pi C_p T_\infty \rho_\infty}. \tag{6.9}$$

Solutions as obtained are called weak plume solutions generated from point sources and are summarized as follows:

$$b = \frac{6\alpha}{5} z, \tag{6.10}$$

$$u = \left(\frac{25}{48} \frac{g}{\pi c_p T_\infty \rho_\infty \alpha^2}\right)^{1/3} \dot{Q}_c^{1/3} z^{-1/3}, \tag{6.11}$$

$$\frac{\Delta \rho}{\rho_\infty} = \frac{5}{6}\left(\frac{10}{9(\pi c_p T_\infty \rho_\infty \alpha^2)^2 g}\right)^{1/3} \dot{Q}_c^{2/3} z^{-5/3}. \tag{6.12}$$

The term $\Delta \rho / \rho_\infty$ can be replaced by $\Delta T / T_\infty$. Generally, the value of α, as experimentally determined is around 0.15. Using this, the temperature difference is expressed as

$$\frac{\Delta T}{T_\infty} = 5\left(\frac{1}{(c_p T_\infty \rho_\infty)^2 g}\right)^{1/3} \dot{Q}_c^{2/3} z^{-5/3}. \tag{6.13}$$

The mass flow rate of the plume as a function of z is calculated as

$$\dot{m}_p = \rho \pi b^2 u. \tag{6.14}$$

Using the assumption that $\rho \approx \rho_\infty$ and $\alpha = 0.15$,

$$\dot{m}_p = 0.2 \left(\frac{\rho_\infty^2 g}{c_p T_\infty} \right)^{1/3} \dot{Q}_c^{1/3} z^{5/3}. \tag{6.15}$$

To improve the ideal plume solutions presented above, the following aspects are considered. The buoyant flow originates from a finite area (fuel bed) that is located at a distance z_0 from a point called virtual origin. Further, the weak plume expressions incorporate uniform temperature and velocity profiles over the flame radius (or width) in the form of top-hat profiles. These are overcome by replacing the top-hat profile by more realistic Gaussian profile as shown in Fig. 6.1. In this case, the plume radius is defined as the distance from the centerline where the temperature has reduced to half of the centerline value at a given height. The plume radius can also be defined on the basis of velocity profile, as the location where the axial velocity has reduced to half of its centerline value. The expressions for plume radius and temperature difference are obtained with the support of experimental measurements. The plume radius (b_T) based on temperature profile is expressed as,

$$b_T = 0.12 \, (T_0/T_\infty)^{0.5} \, (z - z_0). \tag{6.16}$$

Here, T_0 is the centerline temperature at the axial distance of z. The difference between the centerline temperature and the ambient temperature [$\Delta T = T(z) - T_\infty$] is expressed as [2],

$$\Delta T_0 = 9.1 \left(\frac{T_\infty \dot{Q}_c^2}{g c_p^2 \rho_\infty^2} \right)^{1/3} (z - z_0)^{-5/3}. \tag{6.17}$$

The centerline velocity as the function of plume height is expressed as,

$$u_0 = 3.4 \left(\frac{g \dot{Q}_c}{c_p \rho_\infty T_\infty} \right)^{1/3} (z - z_0)^{-1/3}. \tag{6.18}$$

These modified equations are called strong plume relations. In general, these expressions are valid only in the plume region and are not applicable when $z \leq h_f$. The virtual origin (z_0) can be estimated easily when the centerline temperature as a function of z is known accurately. As per the strong plume relation for ΔT_0, a plot of $(\Delta T_0)^{-3/5}$ as a function of z is a straight line, which intersects the z-axis at z_0. A correlation for z_0, following Hesteskad [2], is written as follows:

$$z_0 = 0.083 \dot{Q}^{2/5} - 1.02 D. \tag{6.19}$$

Table 6.1 Correlations for excess temperature (ΔT), centerline velocity (u_0), and virtual origin (z_0)

	ΔT (K)	u_0 (m/s)	z_0 (m)	Refs.
1	$13.74\left[\frac{\dot{Q}_c^{2/5}}{z}\right]^{5/3}$	$0.59\left[\frac{\dot{Q}_c}{z}\right]^{1/3}$	–	[7]
2	$25\left[\frac{\dot{Q}_c^{2/5}}{z-z_0}\right]^{5/3}$	$1.0\left[\frac{\dot{Q}_c}{z-z_0}\right]^{1/3}$	$0.083\dot{Q}^{2/5}-1.02D$	[2]
3	$22.3\left[\frac{\dot{Q}^{2/5}}{z}\right]^{5/3}$	$1.1\left[\frac{\dot{Q}}{z}\right]^{1/3}$	–	[8]
4	$4.58\left[\frac{\dot{Q}_c^{2/5}}{z^*-z_0}\right]^{5/3}$	$1.29\left[\frac{\dot{Q}_c}{z^*-z_0}\right]^{1/3}$	$0.095\dot{Q}^{2/5}-z_{00}$	[9]

z, z^* (m)—distance measured from top of fuel
$\dot{Q}$—total heat release rate (kW), $\dot{Q}_c$ (kW) $\sim 60-80\%$ of $\dot{Q}$
$z_{00} = 1.6$ m (2 tier), 2.4 m (3 and 4 tier)

It should be noted that the total energy release rate (in kW) is used in Eq. (6.19) like in the calculation of flame height expressed in Eq. (6.1). It should also be noted that there are several other correlations available for ΔT, u_0 and z_0, and h_f in literature (Table 6.1). Correlations 1 and 3 in Table 6.1 do not use virtual origin. Correlation 1 is by Zukoski et al. [7] and is based on experiments using the ideal plume theory. Correlation 3 is by McCaffrey who divided the plume into the three zones (flame, intermittent, and plume) and derived separate relations in each zone [8]. Table 6.1 shows the correlation for the plume zone. Correlations 2 and 4 are from work performed by Heskestad [2] and Kung et al. [9]. Heskestad was the first to introduce the concept of virtual origin and relaxed the assumption of a point source necessary for the ideal plume case. Kung et al. [9] performed several experiments on two-tier and four-tier warehouse commodities and modified the empirical constants used by Heskestad [2] to include a correlation for large fires in three-dimensional geometries like in a warehouse storage with two to four tiers of racks filled with boxes. Moreover, in correlation 4, the diameter of the pool is replaced by an average height z_{00} as shown in Table 6.1.

6.3 Air Entrainment Rate

In the derivation shown in Sect. 6.2 for idealized fire plumes, the air entrainment velocity into the plume was assumed to be equal to αu (Fig. 6.2). However, experiments have shown that the air entrainment velocity of weak plumes equals a non-dimensional constant multiplied by the plume velocity at the given height.

Therefore, the air entrainment rate is expressed as,

$$\dot{m}_{ent} = E\rho_\infty u_o (b_T)^2.$$

For strong plumes, by setting $T_0/T_\infty \approx 1$, in Eq. (6.16) written in terms of b_T, and using Eq. (6.18), an expression for air entrainment rate can be obtained as,

$$\dot{m}_{ent} = E\left(\frac{g\rho_\infty^2 \dot{Q}_c}{c_p T_\infty}\right)^{1/3} (z - z_0)^{5/3}. \tag{6.20}$$

Rouse et al. [10] reported a value of E as 0.153. Later, Cetegen et al. [11], assuming that the convective heat release rate as 70% of the total heat release rate, reported the value of E as 0.24. This correlation has been seen to be valid at larger heights ($z > L_f$). Heskestad [12] added one more term to the mass entrainment expression as follows:

$$\dot{m}_{ent} = E\left(\frac{g\rho_\infty^2 \dot{Q}_c}{c_p T_\infty}\right)^{1/3} (z - z_0)^{5/3} \left[1 + \frac{G\dot{Q}_c^{2/3}}{(g^{0.5} c_p \rho_\infty T_\infty)^{2/3} (z - z_0)^{5/3}}\right]. \tag{6.21}$$

It has been shown that $E = 0.196$ and $G = 2.9$ are recommended values to estimate the mass entrainment rates at and above the mean flame height. Under atmospheric conditions, using the above values of E and G, correlation for mass entrainment rate above the mean flame length is given by,

$$\dot{m}_{ent} = 0.071\dot{Q}_c^{1/3}(z - z_0)^{5/3}(1 + 0.027\dot{Q}_c^{2/3}(z - z_0)^{-5/3}). \tag{6.22}$$

At the mean flame length, the correlation is expressed by considering ΔT as 500 K, given as,

$$\dot{m}_{ent} = 0.0056\dot{Q}_c. \tag{6.23}$$

For fire diameters of 0.3 m or greater, the mass entrainment rate for $z \le L_f$ is written as,

$$\dot{m}_{ent} = 0.0059\dot{Q}_c(z/L_f). \tag{6.24}$$

In Eqs. (6.20)–(6.24), the convective heat release rate is in kW, length dimensions are in m, and mass entrainment rate is in kg/s. In general, fires with much lower L_f/D ratios (smaller than 0.9) have not been studied in literature and Eqs. (6.22)–(6.24) have to be used with caution for such cases.

Flame pulsations result because of air entrainment and these have been quantified by several experiments. Cetegen and Ahmed [13] researched on fire diameters in the range of 0.03–20 m and arrived at the correlation for flame pulsation frequency, f (in Hz), in terms of the fire diameter (in m) given in Eq. (6.25), which shows that as D increases, the flame pulsation frequency decreases.

$$f = 1.5/D^{0.5}. \tag{6.25}$$

6.4 Plume Interaction with Ceiling

Several researchers have investigated the effects of the presence of a wall and a corner adjacent to the fire source on the fire plume characteristics. They have reported effects due to the interaction between the plumes and the wall or corner, which are relatively insignificant compared to the interaction of the plume with the ceiling. The plume rises vertically, impinges on the ceiling, and bends in a perpendicular direction. The movement of the plume after impinging with the ceiling is called a *ceiling jet*. The characteristics of the ceiling jet, such as the magnitudes of its temperature and velocity, are quite important for the performance of fire safety devices, such as smoke detectors, heat detectors, and sprinklers, which are usually embedded just below the ceiling. A schematic of a steady ceiling jet flow is shown in Fig. 6.3. When the steady flame height is much smaller than the ceiling height, H, the resulting plume is considered to be a *weak plume*. The ceiling jet thickness, L_T (as shown in Fig. 6.3), at any radial location, r, from the impinging centerline is correlated by Alpert [14] as,

$$\frac{L_T}{H} = 0.112\left[1 - \exp\left(-2.24\frac{r}{H}\right)\right]. \tag{6.26}$$

The locations of the maximum temperature and velocity are dependent on the scale of the fire. Correlations are proposed by Alpert [14] based on experiments involving heat release rate in the range of 668 kW–98 MW and ceiling height in the range of 4.6–15.5 m. These are expressed as,

$$\begin{aligned} T - T_\infty &= 16.9\dot{Q}^{2/3}/H^{5/3} \quad \text{for } r/H \le 0.18 \\ T - T_\infty &= 5.38\frac{\dot{Q}^{2/3}/H^{5/3}}{(r/H)^{2/3}} \quad \text{for } r/H > 0.18 \end{aligned} \tag{6.27}$$

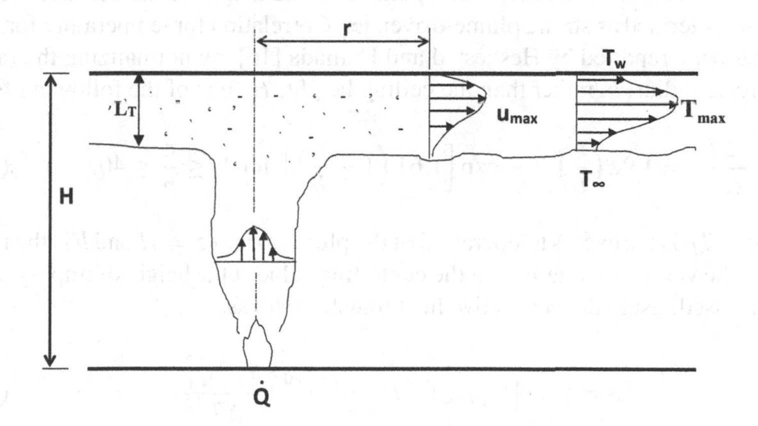

Fig. 6.3 Schematic of a ceiling jet

$$u = 0.96(\dot{Q}/H)^{1/3} \quad \text{for r/H} \le 0.15$$
$$u = 0.195\frac{(\dot{Q}/H)^{1/3}}{(r/H)^{5/6}} \quad \text{for } r/H > 0.15. \tag{6.28}$$

In the correlations in Eqs. (6.27) and (6.28), the total heat release is expressed in kW, r and H are in m, and u in m/s. It is clear that the ceiling jet temperature and velocity near to the plume impingement region, called the *turning region*, do not depend on the value of radius. Outside the turning region, the correlations are expressed in terms of radius. These correlations are generally applied to a time period after fire ignition, when the hot gas accumulation in the ceiling is not high and the fire is not affected by radiation from the walls. For near-wall or near-corner fires, these correlations have to be modified. The general practice is to multiply the heat release by 2 for near-wall fire and by 4 for corner fire and use it in these correlations.

Heskestad [15] reported non-dimensional correlations as follows:

$$\dot{Q}_0^* = \frac{\dot{Q}}{\rho_\infty c_p T_\infty g^{0.5} H^{2.5}}; \; \Delta T_0^* = \frac{\Delta T/T_\infty}{(\dot{Q}_0^*)^{2/3}}; \; u_0^* = \frac{u/\sqrt{gH}}{(\dot{Q}_0^*)^{1/3}}. \tag{6.29}$$

Combining Alpert's theory and Heskestad's correlation, the following expressions are obtained:

$$\Delta T_0^* = \left(0.225 + 0.27\frac{r}{H}\right)^{-4/3} \quad \text{for } 0.2 \le \text{r/H} \le 4.0$$
$$\Delta T_0^* = 6.3 \text{ for r/H} \le 0.2 \tag{6.30}$$

$$u = 1.06\left(\frac{r}{H}\right)^{-0.69} \quad \text{for } 0.17 \le \text{r/H} \le 4.0$$
$$u = 3.61 \text{ for r/H} \le 0.17. \tag{6.31}$$

When the height of the fire is comparable to the height of the ceiling, then the ceiling jet is termed as strong plume-driven jet. Correlation for temperature for these cases has been reported by Heskestad and Hamada [16], by normalizing the radius, r, by plume radius, b, rather than the ceiling height, H. It is of the following form:

$$\frac{\Delta T}{\Delta T_P} = 1.92\left(\frac{r}{b}\right)^{-1} - \exp\left[1.61\left(1 - \frac{r}{b}\right)\right] \text{ for } 1 \le \frac{r}{b} \le 40. \tag{6.32}$$

Here, ΔT_P is the excess temperature of the plume axis at $z = H$ and b is the radius at which the velocity is one-half of the centerline value at the height of impingement. It is expressed, using the convective heat transfer rate, as,

$$b = 0.42\left[(c_p\rho_\infty)^{0.8} T_\infty^{0.6} g^{0.4}\right]^{-0.5} \frac{T_P^{0.5}\dot{Q}_C^{2/5}}{\Delta T_P^{0.6}}, \tag{6.33}$$

where T_P is the centerline temperature at the height of ceiling.

The correlations discussed so far are applicable to steady fires. These can be used in the case of slowly growing unsteady fires, by replacing the constant heat release rate to time-dependent heat release rate. For instance, for pre-flashover compartment fire (dealt in Chap. 7) or a growing fire, the heat release rate is expressed as $\alpha_m t^2$, where α_m depends on material properties. Further, these correlations for unconfined ceiling jets are applicable for large storage facilities and industries. In a typical smaller compartment fire, these correlations cannot be applied without modifications. In smaller compartments or long duration fires, the smoke layer accumulates under the ceiling and completely submerge the ceiling jet. Correlations for convective heat transfer from plumes and ceiling jets to ceilings are reported by several researchers in literature. Similarly, researchers have worked on ceiling jets over sloped ceilings and have proposed correlations for temperature and velocity as a function of slope angle.

6.5 Case Studies

Case Study 1: Puffing Methanol Pool Flame A lab-scale experiment has been conducted with 6 cm pool burner using methanol as the fuel. Figure 6.4 illustrates the instantaneous direct photographs of flame images at several time instants.

Methanol flames are non-luminous as well as non-sooty. The visible flame height for each image is reported beneath the image. It is clear that the flame is oscillatory beyond a certain height as a result of instabilities caused due to buoyancy induced flow. From the series of photographs, it can be concluded that the combustion zone prevails up to a height of around 8 cm and beyond that an intermittent zone prevails. Also, the ratio of the maximum flame length to the minimum flame length is little more than 2 as reported in literature. The average mass burning rate is measured using a load cell as 0.038 g/s. The average heat release rate is around 0.7562 kW, as calculated using the lower calorific value of methanol. The average flame length calculated using Eq. 6.2 is 0.144 m or 14.4 cm, which is a good estimate in comparison

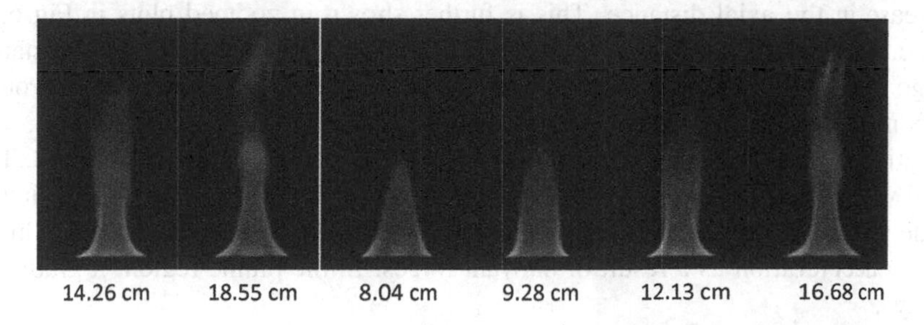

Fig. 6.4 Direct photographs of six instantaneous methanol pool flames showing the flame height, diameter of the pool = 6 cm

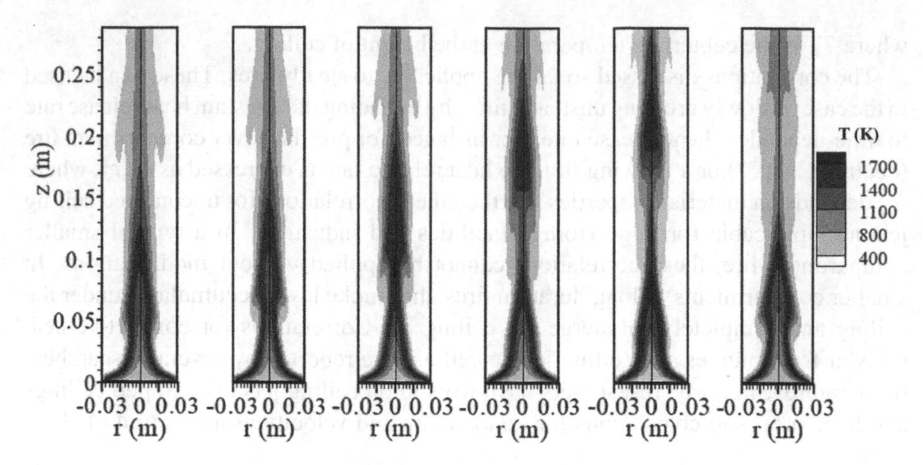

Fig. 6.5 Instantaneous grayscale temperature contours of methanol pool flames

to the average visible flame height calculated from six instantaneous images shown in Fig. 6.4, which is equal to 13.2 cm.

Case Study 2: Plume Characteristics In this case study, transient numerical simulation of unconfined burning of 50 mm diameter methanol pool is presented. An axisymmetric domain of diameter 100 mm and height of 1.5 m has been chosen for simulating the methanol pool flame. Ansys FLUENT has been used in the simulations. The interface boundary conditions for steady burning of liquid fuel pool, as discussed in Chap. 4, have been implemented using a User-Defined Function. A global single step reaction is used to model chemical kinetics. Figure 6.5 shows the grayscale temperature contours at several time instants, indicating the fluctuations in the temperature field in the reaction and intermittent zones.

The elongation of the flame, its puffing, and shedding is shown in Fig. 6.5. The reaction and intermittent zones extend up to 0.3 m, which is around six times the pool diameter. The axial variation of time-averaged temperature is shown in Fig. 6.6a–c. The variation along the whole domain length of 1.5 m is shown in Fig. 6.6a. The temperature increases along the axis, reaches a maximum, and decays with further increase in the axial distance. This is further shown in zoomed plots in Fig. 6.6b and c. The temperature increases and remains almost constant in the axial distance range of 0.075–0.1 m. It then decreases gradually up to an axial distance of around 0.25 m. After this, it decreases in a non-linear manner in the plume region. The variation of time-averaged axial velocity along the axis is shown in Fig. 6.6d. The axial velocity increases from a low value from the pool surface to reach a maximum value around 0.3 m, up to which the reaction and intermittent zones extend. This is due to acceleration as a result of buoyant forces. In the plume region, it gradually decays.

Radial time-averaged temperature profiles at several axial locations are shown in Fig. 6.7a. Temperature decreases from its maximum value at the axis to attain the ambient temperature asymptotically, following a typical Gaussian profile. As the

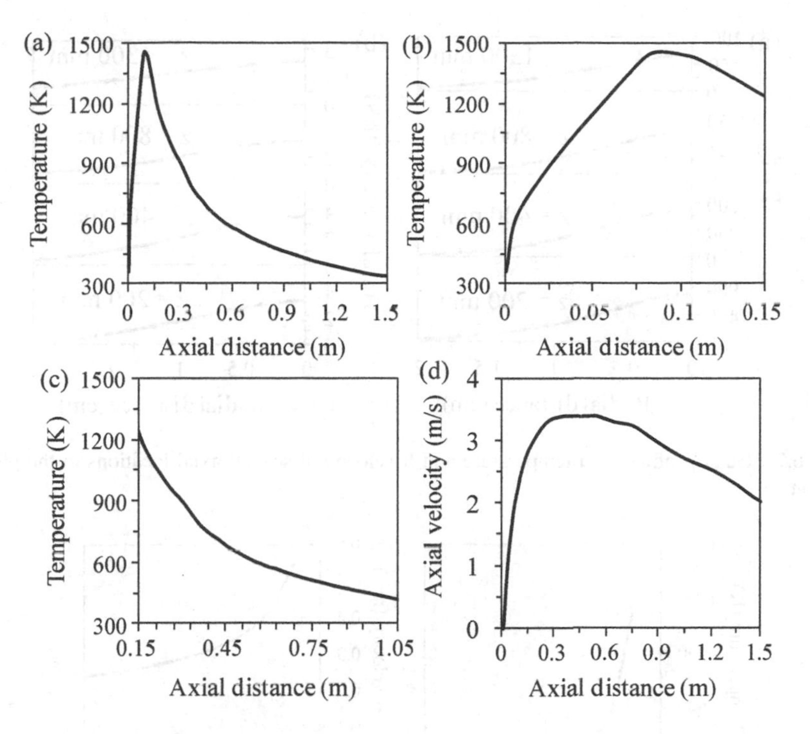

Fig. 6.6 Variation of temperature and axial velocity along the axis

axial location increases, the maximum temperature value decreases (Fig. 6.6) as a result of entrainment of cooler ambient air. The radial profiles of time-averaged axial velocity at several axial locations are shown in Fig. 6.7b. The profiles are similar to those of the temperature; the value is the maximum at the axis and it decreases along the radial direction.

The maximum value of axial velocity also decreases as z increases. Moreover, the radial velocity profiles are flatter at higher z-locations. The plume radius can be evaluated from these radial profiles. Plume radius based on the temperature is evaluated as the radial location where the temperature at the axis decreases to half of its value. Similarly, the plume radius based on the axial velocity is evaluated as that radial location where the axial velocity has decreased to half of its centerline value. The plume radius increases with an increase in the axial location.

The centerline temperature and velocity profiles calculated using Eqs. (6.17) and (6.18) are shown in Fig. 6.8a and b, respectively. The heat release rate ($\dot{Q}_c$) has been evaluated using mass loss rate of methanol using B number equation (refer Chap. 5).

For the heat release rate of 12.4 kW, the flame height is calculated using Eq. (6.19) as 17.6 cm. The temperature values are seen to be comparable to numerical predictions. However, the predicted axial velocity using Eq. (6.18) is much lower than the numerical simulations. This is because of the simplifications needed to arrive at Eq. (6.18).

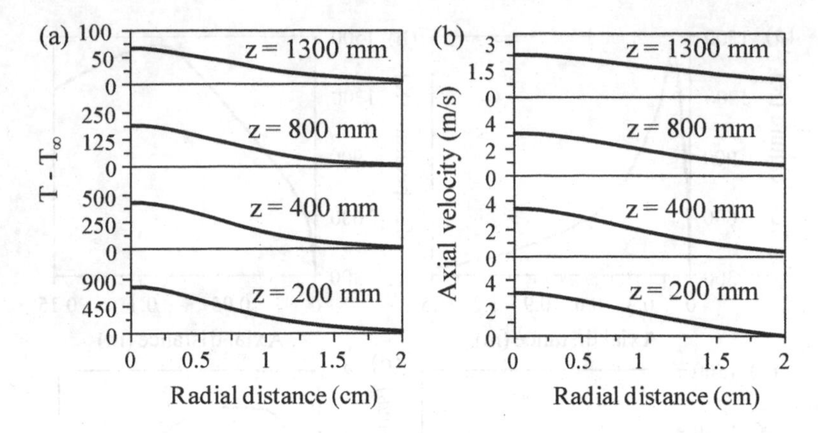

Fig. 6.7 Radial profiles of **a** temperature and **b** velocity at several axial locations in the plume region

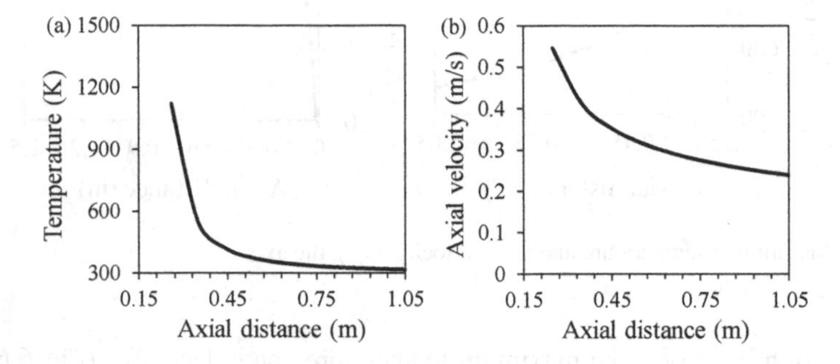

Fig. 6.8 Variation of temperature and axial velocity along the axis using Eqs. (6.17) and (6.18)

Case Study 3: Ceiling Jet Characteristics For understanding the characteristics of ceiling jets, lab-scale flames from MMA pool have been simulated using Ansys FLUENT. The diameter of the MMA pool is 25 mm. An axisymmetric domain is used where the ceiling height (H) (location of the horizontal obstruction on the top) is varied from 6 to 24 cm. The radius of the ceiling has been kept as 7.5 cm.

Figure 6.9 presents the grayscale temperature contours and velocity vectors for the cases with different ceiling heights. It is clear from Fig. 6.10 that strength of the ceiling jet depends on the plume strength and it decreases as the ceiling height increases (or the plume becomes weaker). The temperature reached by the ceiling decreases as the ceiling height increases. The distance from the ceiling, where the temperature drops to nearly the ambient value, decreases as the ceiling height is increased.

As far as the ceiling jet velocity is concerned, at a radial location near to the impingement point, the maximum velocity increases with an increase in the ceiling height and the location of the maximum velocity from the ceiling decreases. At farther radial locations, the differences in the values of the maximum velocity decrease. The

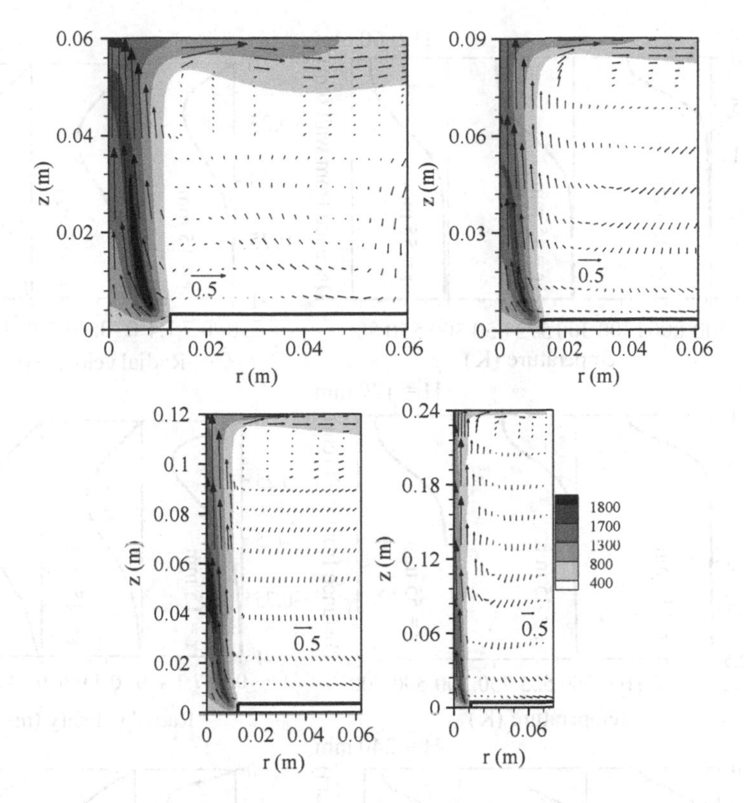

Fig. 6.9 Grayscale temperature (in K) contours and velocity (in m/s) vectors for cases with varying ceiling heights

location of the maximum velocity from the ceiling decreases with increasing ceiling height. Therefore, it is clearly illustrated that the plume and ceiling jet dynamics depend on heat release rate and the ceiling height. Ceiling jet is further affected by the ceiling vents and due to the presence of obstructions such as beams present in the ceiling.

Case Study 4: McCaffrey Correlations Consider an 8 m high (W24 × 120) steel column in a pool fire that can support its structural load provided the steel temperature remains below 540 °C. The column is set in a normal weight concrete footing (0.7 m thick) as shown in Fig. 6.11. The objective of the case study is to provide a solution strategy for determining the largest diameter of a flammable liquid pool fire that the column can withstand for an indefinitely long duration and the corresponding heat release rate of the fire. The following assumptions have been used:

1. The convective heat transfer coefficient, (h_c), is a constant = 5 W/m^2-K.
2. Emissivity = 0.7 for radiation from flame, intermittent, and plume zones.
3. Ambient temperature = 20 °C.

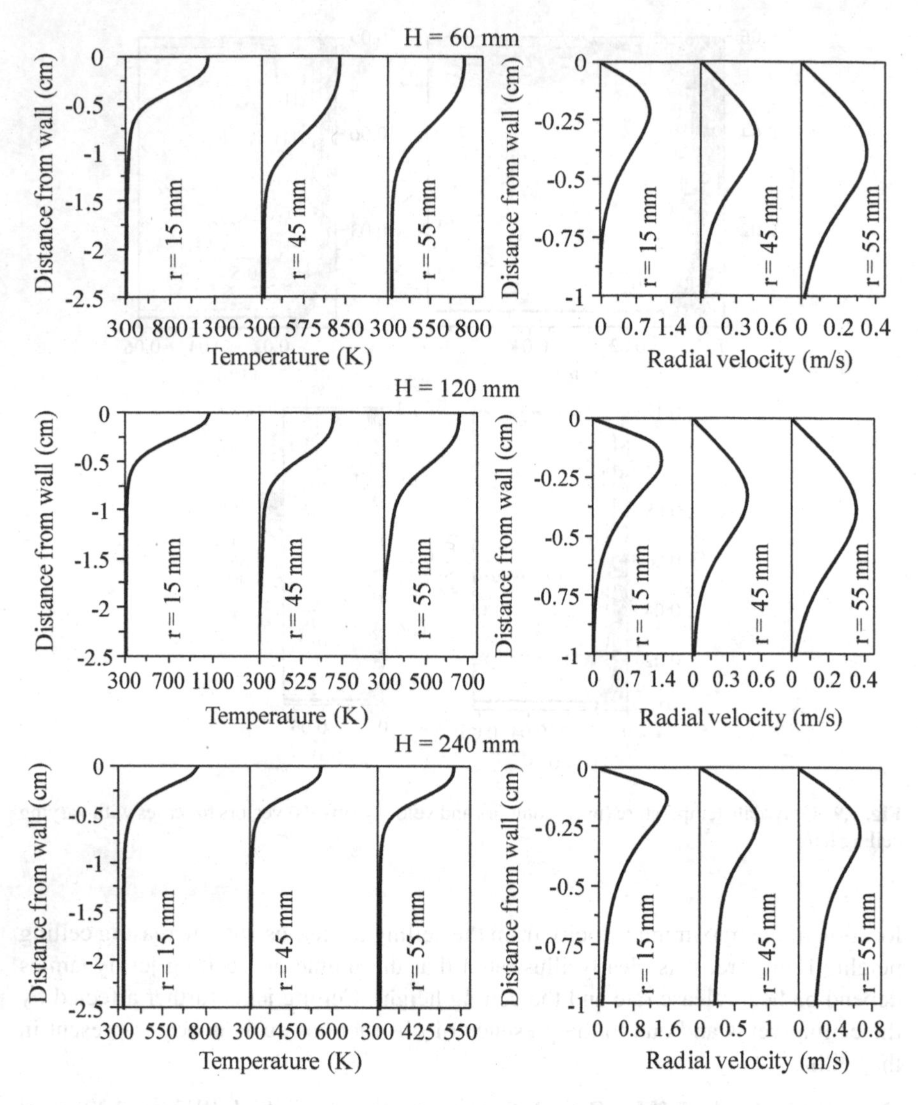

Fig. 6.10 Profiles of temperature and velocity at different radial positions

4. The conduction loss in the concrete is linear, thermal properties of concrete do not change with temperature, and the concrete is surrounded by soil which ensures a constant temperature to the outer wall of the concrete equal to 20 °C.
5. The HRR of a pool fire can be approximated as 1.5 MW/m^2.

Given these parameters, the correlations for centerline plume temperature from McCaffrey [8], which are shown in Table 6.2, can be used, where $\dot{Q}$ is in kW and z is

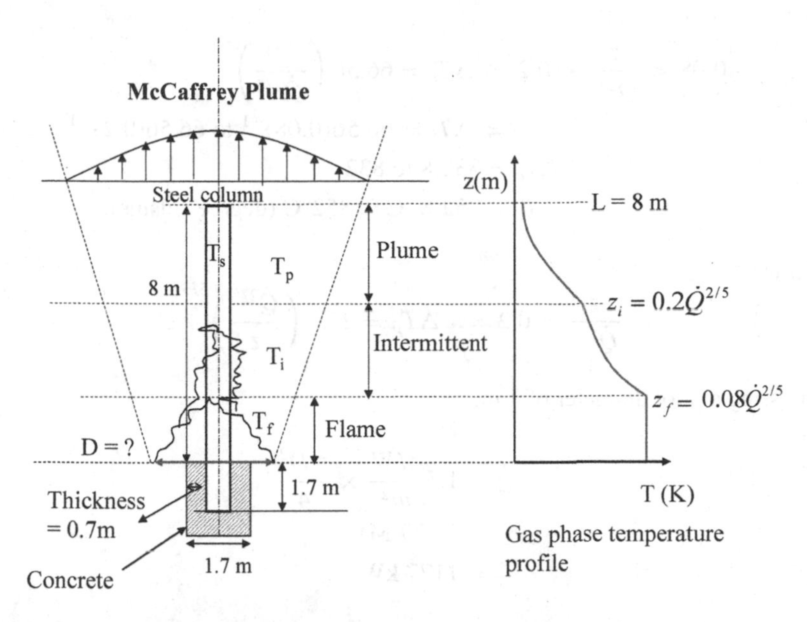

Fig. 6.11 Schematic of a steel column in a pool fire

Table 6.2 Correlations for excess temperature (ΔT in K) in the flame, intermittent, and plume regions

$$\Delta T = \frac{T_\infty}{2g}\left(\frac{k}{0.9}\right)^2\left(\frac{z}{\dot{Q}^{2/5}}\right)^{2\eta-1}$$

Zone	$\frac{z}{\dot{Q}^{2/5}}$ (m/kW$^{2/5}$)	k	η
Flame	<0.08	6.8 m$^{1/2}$/s	1/2
Intermittent	0.08–0.2	1.9 m/(s-kW$^{1/5}$)	0
Plume	>0.2	1.1 m$^{4/3}$/(s-kW$^{1/3}$)	−1/3

in m. The three zones are denoted using subscripts f, i, and p for flame, intermittent, and plume regions, respectively, in the formulation given below:

- Flame:

$$\frac{z}{\dot{Q}^{2/5}} < 0.08 \Rightarrow \Delta T_f = 853 \text{ or } T_f = 873\,^\circ\text{C}.$$

- Intermittent:

$$0.08 < \frac{z}{\dot{Q}^{2/5}} < 0.2 \Rightarrow \Delta T_i = 66.56\left(\frac{z}{\dot{Q}^{2/5}}\right)^{-1}$$

$$\Rightarrow \Delta T_i = 66.56(0.08)^{-1} \text{ to } 66.56(0.2)^{-1}$$

$$\Delta T_i = 332.8 \text{ to } 832$$

$$T_i = 352.8\,°\text{C to } 852°\text{C (degree celsius).}$$

- Plume:

$$\frac{z}{\dot{Q}^{2/5}} > 0.2 => \Delta T_p = 22.3\left(\frac{\dot{Q}^{2/5}}{z}\right)^{5/3}.$$

Guessing a pool diameter of 1 m,

$$\dot{Q} = 1.5\frac{MW}{m^2} \times \frac{\pi D^2}{4}$$
$$= 1.177 \text{ MW}$$
$$= 1177 \text{ kW}.$$

Flame height is evaluated as $z_f = 0.08\dot{Q}^{2/5} = 1.35$ m. The z-location of the intermittent zone is given by $z_i = 0.2\dot{Q}^{2/5} = 3.4$ m.
It is interesting to note that the Heskestad flame height as evaluated using Eq. (6.2) is found to be $0.23(1177)^{2/5} - 1.02 = 2.87$ m which lies in between z_f and z_i.

The average temperature in the three zones can be evaluated using following expressions:

$$\overline{T_f} = T_f$$
$$= 1146\ K$$

$$\overline{T_i} = \int_{z_f}^{z_i} \frac{\Delta T}{(z_i - z_f)} dz$$
$$= \int_{z_f}^{z_i} \frac{66.56}{(z_i - z_f)}\left(\frac{z}{\dot{Q}^{2/5}}\right)^{-1} dz$$
$$= 800.3 \text{ K}$$

$$\overline{T_p} = \int_{z_i}^{L} \frac{\Delta T}{(L - z_i)} dz$$
$$= \int_{z_i}^{L} \frac{22.3}{(L - z_i)}\left(\frac{z}{\dot{Q}^{2/5}}\right)^{5/3} dz$$
$$= 442 \text{ K}.$$

For conservation of energy,

$$\Sigma \dot{q}'' \times A = 0$$

where $A = \pi DL$, where (πD) is a constant. Accounting for the heat transfer in flame, intermittent, and plume zones and the heat transfer to the concrete base:

$$\Sigma \dot{q}'' \times L = 0 => (h_{T,f})(\overline{T_f} - T_s) \times z_f + (h_{T,i})(\overline{T_i} - T_s) \times (z_i - z_f)$$

$$+(h_{T,p})(\overline{T_p} - T_s) \times (L - z_i) - k_c \frac{(T_s - T_\infty)}{\delta} \times 1.7 = 0,$$

where h_T represents the summation of convective heat transfer coefficient (h_c) and linearized radiation heat transfer coefficient (h_r) given as:

$$\begin{aligned} h_T &= h_c + h_r \\ &= h_c + (\overline{T} + T_s)((\overline{T}^2 + T_s^2) \times \sigma\epsilon), \end{aligned}$$

where $\overline{T}$ represents average temperature in the zone, T_s is the steel column temperature, and $\sigma = 5.67 \times 10^{-8}$ W/m^2-K^4 is the Stefan-Boltzmann constant. Using h_c = 5 W/m^2-K and ϵ = 0.7,

$$\begin{aligned} h_{T,f} &= 5 + (1146 + 540 + 273)(1146^2 + 813^2) \times \sigma\epsilon \\ &= 5 + 153.5 = 158.5 \text{ W/m}^2\text{-K} \\ h_{T,i} &= 5 + (800.3 + 813)(800.3^2 + 813^2) \times \sigma\epsilon \\ &= 5 + 83.33 = 88.33 \text{ W/m}^2\text{-K} \\ h_{T,p} &= 5 + (442 + 813)(442^2 + 813^2) \times \sigma\epsilon \\ &= 5 + 42.65 = 47.65 \text{ W/m}^2\text{-K}. \end{aligned}$$

It is seen here that the linearized radiative heat transfer coefficient (h_r) is higher (153.5 W/m^2-K, 83.33 W/m^2-K, and 42.65 W/m^2-K in flame, intermittent, and plume regions, respectively) than the convective heat transfer coefficient (5 W/m^2-K). Thus, radiation is the dominant mode of heat transfer in this case. For conservation of energy and using thermal conductivity of concrete = 1.6 W/m-K, we get

$$\begin{aligned} \Sigma \dot{q}'' \times L &= 158.5 \times (1146 - 813) \\ &\quad \times 1.35 + 88.33 \times (800.3 - 813) \times 2.05 \\ &\quad + 47.65 \times (442 - 813) \times 4.6 - 1.6 \\ &\quad \times (813 - 293) \times 1.7/0.7 \\ \Sigma \dot{q}'' \times L &= 71.25 - 2.3 - 81.32 - 2.02 \\ &= -14.39 \text{ kW/m}. \end{aligned}$$

This procedure can be repeated iteratively until $\Sigma \dot{q}'' \times L = 0$ by varying the pool diameter. It should be noted that the natural convection heat transfer coefficient usually varies between 5 and 15 W/m^2-K. It will not be a constant along the length of the column as assumed in this case study. One way of estimating h_c would be to use Nusselt number correlations, which are available in standard heat transfer textbooks. h_c may be evaluated for the three zones independently using the Nusselt number correlation.

Review Questions

1. What is buoyancy force?
2. Write the governing equations involving plume transport?
3. What are the parameters affecting the temperature of the plume?
4. What are the variables involved in the calculation of mass flow rate of the plume?
5. Explain the term half-width of the plume with suitable expressions.
6. Draw typical temperature and velocity profiles at a given axial location of an axisymmetric plume.
7. A hot jet of air at 1200 K is injected into a quiescent atmosphere at 300 K. The temperature decays as $T = 1200 - 300z - 150z^2, 0 \leq z \leq 2$ (z is in m). Estimate the velocity of plume at 0.8 m from the source.
8. How is air entrainment into the plume calculated?
9. LPG (70% butane, 30% propane by volume) is supplied with a velocity of 1.6 cm/s from a circular burner of diameter 10 mm. Estimate the height of the flame and the air entrainment velocity (assume stoichiometric combustion).
10. Hot combustible gases at 1500 K (molecular weight = 29 kg/kmol) is ejected into a quiescent ambient air (300 K) from a circular port of diameter 50 mm at 2 m/s. Calculate the radius of the plume at 0.5 m from the source. Also, evaluate the convective heat transfer in the plume and mass flow rate of the plume if the specific heat of gases is 1.5 kJ/kg-K.
11. Explain the factors influencing the transport of a ceiling jet.
12. What are weak and strong plumes?
13. Draw typical temperature and velocity profiles of a ceiling jet.
14. A combustible material is burnt in a closed room of height 5 m and length 7 m. The material is located at 5 m from the exit. The heat release from the flame is 800 kW. Estimate the thickness of the ceiling jet at 2 m from the exit. Also calculate the velocity and temperature of the plume at this location.

References

1. Quintiere, J.G. 2017. *Principles of Fire Behavior*. Boca Raton: CRC Press.
2. Heskestad, G. 1995. *S.F.P.E. Handbook, and of Fire Protection Engineering*, 2nd ed. Quincy: National Fire Protection Association.

3. McCaffrey, B. 1995. *S.F.P.E. Handbook, and of Fire Protection Engineering*, 2nd ed. Quincy: National Fire Protection Association.
4. Heskestad, G. 1983. Luminous heights and turbulent diffusion flames. *Fire Safety Journal* 5: 103–108.
5. Karlsson, B., and J.G. Quintiere. 1999. *Enclosure Fire Dynamics*. Boca Raton: CRC Press.
6. Morton, B.R., G.I. Taylor, and J.S. Turner. 1956. Turbulent gravitational convection from maintained and instantaneous sources. *Proceedings of the Royal Society of London* A234: 1–23.
7. Zukoski, E.E., T. Kubota, and B. Cetegen. 1981. Entrainment in fire plumes. *Fire Safety Journal* 3 (2): 107–121.
8. McCaffrey, B. 1979. *Purely Buoyant Diffusion Flames: Some Experimental Results*. National Bureau of Standards (USA). NBSIR 79–1910.
9. Kung, H.C., H.Z. You, and R.D. Spalding. 1988. Ceiling flows of growing rack storage fires. *Proceedings of Combustion Institute* 21: 121–128.
10. Rouse, H., C.S. Yih, and H.W. Humphreys. 1952. Gravitational convection from a boundary source. *Tellus* 4 (3): 201–210.
11. Cetegen, B.M., E.E. Zukoski, and T. Kubota. 1984. Entrainment in the near and far field of fire plumes. *Combustion Science and Technology* 39 (1–6): 305–331.
12. Heskestad, G. 1986. Fire plume air entrainment according to two competing assumptions. *Proceedings of Combustion Institute* 21 (1): 111–120.
13. Cetegen, B.M., and T.A. Ahmed. 1993. Experiments on the periodic instability of buoyant plumes and pool fires. *Combustion and Flame* 93 (1–2): 157–184.
14. Alpert, R.L. 2016. *Ceiling Jet Flows, S.F.P.E. Handbook, and of Fire Protection Engineering*, vol. 5, 429–454. New York: Springer.
15. Heskestad, G. 1975. Physical modeling of fire. *Journal of Fire and Flammability* 6: 253–273.
16. Heskestad, G., and T. Hamada. 1993. Ceiling jets of strong fire plumes. *Fire Safety Journal* 21 (1): 69–82.

Chapter 7
Compartment Fires

A compartment fire is a fire in an enclosed space and involves understanding ignition of materials, pyrolysis, flame dynamics, plumes, hot layers, wall heat transfer, and fire-induced flows through openings such as vents, doors, and windows. The enclosed space can be a small closet; single room in a building; a large atrium; a vehicle such as a car, an aircraft cabin, or a space shuttle. The goal from a fire safety perspective is to calculate the response of the fire on the structure and people. The solution is usually an understanding of the evolution of the heat release rate in the compartment and a prediction of what will happen if there is a change, for example, to an opening.

7.1 Stages in a Compartment Fire

The development of a compartment fire is shown in Fig. 7.1 where a room (2.4 m × 2.4 m × 3.7 m or 8′ × 8′ × 12′) made up of gypsum wall board filled with furniture typically found in a student dorm is set on fire. One wall of the compartment has been removed and serves as an opening for flow in and out of the compartment. Ignition is by means of a waste paper basket that was filled with newspaper.

As shown in Fig. 7.1, several stages occur during the development of the fire:

1. **Ignition**: Ignition is a process that produces an exothermic reaction characterized by an increase in temperature significantly higher than ambient. It can occur by piloted ignition (flame, spark, hot spot, high radiative heat flux) or by spontaneous ignition. In the dorm room experiment, ignition was achieved by lighting of the

A. S. Rangwala and V. Raghavan, *Mechanism of Fires*,
https://doi.org/10.1007/978-3-030-75498-3_7

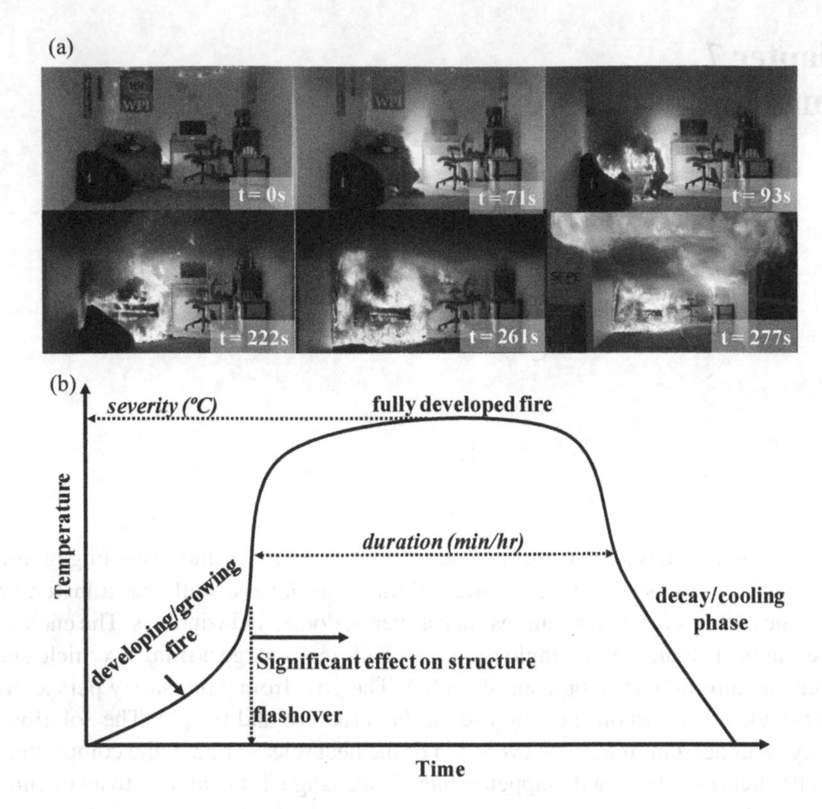

Fig. 7.1 **a** Stages in a typical compartment fire. Ignition, ignition of first item, ignition of second item, flames touch ceiling, development of thick smoke layer, fully developed compartment fire, and **b** the life cycle of a typical compartment fire

waste paper in the trash bin located close to the bed by a propane torch ($t = 0$ s). This results in a flaming combustion scenario as shown in Fig. 7.1a.

2. **Growth**: The fire grows within the basket and soon ignites the bedding at $t = 71$ s as shown in Fig. 7.1. In the early stages, the enclosure seems to have no effect on the fire. The fire spreads and ignites the pillow on the bed and the adjacent futon at $t = 93$ s. The flame heights eventually touch the ceiling at $t = 222$ s, and fire spreads to other objects such as the office chair and side table at $t = 222$ s. The hot gases rise upward because of buoyancy and fill the ceiling as a light gray smoke initially ($t = 93$ s) and then as thick black smoke. The smoke layer is clearly seen at $t = 261$ s. The temperature of the smoke layer varies between 500 and 600 °C and is highly effective in radiatively heating the entire enclosure. As the fire grows in size, the smoke layer descends and increases in temperature. It should be noted here that only a small portion of the actual fuel mass is in the smoke layer formed below the ceiling. The smoke layer is formed because of air entrainment arising because of the temperature gradients between the hot products of combustion and the surrounding cold air. Figure 7.1b shows the role of the developing, growing

fire in terms of a broader view of the overall life cycle of a compartment fire. This stage is significant from a perspective of smoke movement and safe egress of people, and does not affect the structure yet.

3. **Flashover**: Transition of the fire from a single object burning in the compartment to the sudden growth to its biggest possible size (occurs at $t = 277$ s). Flashover is a thermal instability phenomena between the growth stage and the fully developed fire as shown by the sudden/dramatic change in the slope in Fig. 7.1b. Several definitions of flashover can be found in literature. Some rules of thumb are, for example, temperature in the compartment reaches 500–600 °C, radiation to the floor of the compartment is 15–20 kW/m^2, and flames appear from openings. Flashover forms the link between a developing/growing fire and the fully developed fire.
4. **Fully developed fire**: This is the longest and most dominant stage and is usually used as a basis for fire protection design of buildings because during this stage, the structure is affected by the fire. From a fire protection viewpoint, it then becomes imperative to define/quantify the fire boundary condition which can then be used to determine the thermal response of the structure, structural response of the structure, and the evacuation protocol. Based on the ventilation it is observed that a fully developed fire follows two behaviors—*Ventilation Control* (Regime I) and *Fuel Controlled* (Regime II). In Ventilation Control (Regime I), the fire is oxygen starved, and the smoke flowing out of the compartment has no oxygen whatsoever. Thus, it is likely to burn outside. In Regime II, the fire is fuel starved. In this case, the smoke coming out of the compartment has oxygen, does not have excess fuel left over, and will not burn outside the window. A fully developed fire will burn till there is fuel in the compartment.
5. **Decay**: Natural decay because of lack of fuel or oxygen or artificial decay because of suppression (usually water). During the decay stage, the heat release rate in the compartment reduces. The average temperature also decreases, and usually a value of 80% of its peak value is used to identify this stage. For most structures, the decay usually occurs at a linear rate of 5 °C/min.

For fire protection engineering design purposes, the fully developed fire stage is the most important as the fire reaches maximum temperatures during this stage thereby attaining a worst-case scenario condition. Compartment fires and especially the fully developed stage were extensively studied in the 1960s–early 70s where various research programs Kawagoe (Japan) [1], Harmathy (Canada) [2], Thomas (UK) [3], and Emmons (USA) [4] studied the behavior of fire inside buildings. Various experiments, both small and full scale and mathematical models contributed to the fundamental analysis of the physics governing the fire behavior in compartments. These early studies were also responsible for the development of the temperature versus time curve as shown in Eq. (7.1) used in many standards today (ASTM E119, BS 476, Part 20, and ISO 834) to provide a general description of the fire environment in an enclosure:

$$T - T_0 = 345 \log_{10}(8t + 1), \tag{7.1}$$

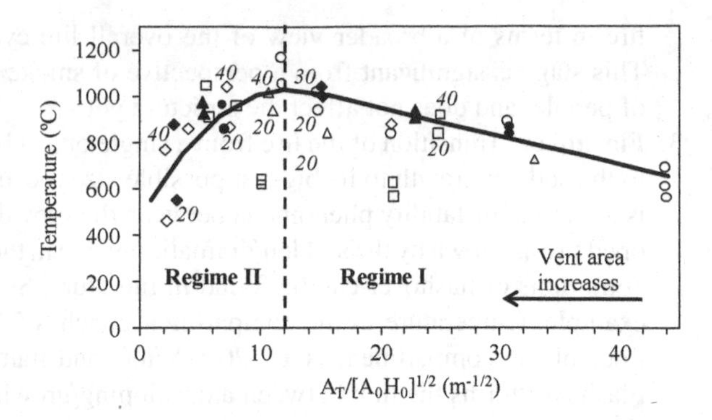

Fig. 7.2 Regime 1 and 2 fires from Thomas and Heseldon's [7] experiments, symbols represent different vent shapes, and the numbers represent the fire load densities in kg/m^2

where T is the temperature of the compartment, T_0 is the initial temperature, and t is the time in seconds.

One of the most comprehensive treatments on the subject of compartment fires is by Karlsson and Quintiere [5] and this chapter is mostly adapted from this classic reference. The compartment fire research has been recently revisited with further insight on the fully developed fuel-controlled stage (Regime II) by Torero et al. [6].

The *duration* and *severity* of a compartment fire are the two main components in understanding the fire behavior during the fully developed stage. The duration is calculated by dividing the total mass of fuel (fuel load) available (kg) by the burning rate of fuel (kg/s). The severity is assessed by the maximum compartment temperature. An important distinction must be made between a fire burning in the open (no compartment) and in a compartment. In a fire burning in the open, the rate of burning induces the rate of air entrainment. This results in air entrainment flows that feed the flame and corresponding growth. However, in a compartment, the rate of burning and the rate of air entrainment need to be considered separately. Figure 7.2 [7] shows two scenarios observed in compartment fires in which all surfaces are burning and hence represent post-flash over or fully developed fire scenarios. A vertical dotted line separates the experimental data points into two regimes that are observed in fully developed fires as shown in Fig. 7.2. The right-hand side region shows maximum compartment temperatures measured experimentally for fully developed fires where flames are coming out of the vent and burning takes place both inside and outside, indicating that the quantity of air available inside the compartment is not sufficient to burn all the fuel generated. Such fires are called ventilation limited (Regime I) fires.

Figure 7.2 shows that such fires follow a linear trend, where the mass burning rate or severity increases linearly with ventilation factor till

$$\frac{A_T}{A_0 H_0^{1/2}} \text{ reaches } \sim 12.$$

Here, A_T represents the area of the compartment excluding the floor and openings, A_0 is the area of the openings, and H_0 is the height of the opening. After this, the burning rate is represented by a Regime II fire. In Regime II fires, the fire scenario is once again fully developed. However, burning is dependent on the quantity of fuel. The *duration* and *severity* of a compartment fire are thus dependent on which regime the burning behavior lies. It will be shown later, that if in Regime I, the burning rate is almost independent of the amount of fuel and its surface area and only proportional to the amount of air supply through the openings. This greatly simplifies the problem and the fire boundary condition in several codes and standards (for example, see SFPE engineering guide to fire exposure to structural elements, 2004) relies on the ventilation factor:

$$\frac{A_T}{A_0 H_0^{1/2}}.$$

7.2 Smoke Filling in a Developing Fire

Figure 7.3 shows a compartment with a fire located at the center of the floor and consequent smoke filling the compartment from the ceiling downwards. Conservation of mass on the smoke layer (CV_1) gives

$$\dot{m}_a + \dot{m}_f = \frac{dm_{CV_1}}{dt}, \tag{7.2}$$

where $\dot{m}_a$ equals the mass of air entrained into the fire, $\dot{m}_f$ is the mass of fuel, and m_{CV_1} is the mass of smoke which accumulates in CV_1 as shown in Fig. 7.3.

The mass of gases entering the plume via combustion is very small compared to the air entrained for buoyant natural convective fires as shown in Chap. 6 or $\dot{m}_f/\dot{m}_a \ll 1$. Further the air entrained into the plume, $\dot{m}_a$, can be obtained using Zukoski's plume entrainment correlation given by,

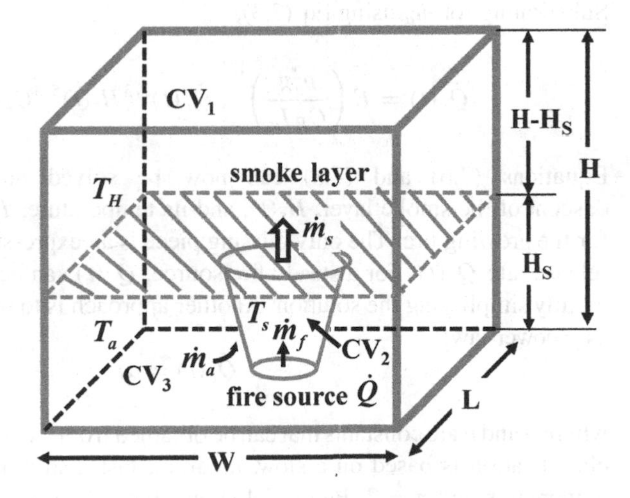

Fig. 7.3 Smoke filling in a compartment

$$\dot{m}_a = E\left(\frac{\rho_a^2 g}{C_P T_a}\right)^{1/3} \dot{Q}_c(t)^{1/3} H_s(t)^{5/3}, \tag{7.3}$$

where $\dot{Q}_c(t)$ is the convective heat release rate, $H_s(t)$ is the smoke layer height, and E is an entrainment constant equal to 0.21 for a fire that is located in the center of the room as shown in Fig. 7.3. m_{CV_1} can be expressed as:

$$m_{CV_1} = \rho_s(L \times W)(H - H_s(t)). \tag{7.4}$$

In Eq. (7.4), ρ_s is the density of the smoke, and L, W, and H are the length, width, and height of the enclosure as shown in Fig. 7.3. As $\dot{m}_F$ is negligible, Eq. (7.2) can be written as:

$$\frac{dm_{CV_1}}{dt} = \dot{m}_a. \tag{7.5}$$

Assuming constant pressure in CV_1 and CV_2 ideal gas law gives $\rho_s = (\rho_a T_a)/T_s(t)$ and substituting into Eq. (7.4) and then into Eq. (7.5) gives a conservation of mass for CV_1 as:

$$(\rho_a T_a)(L \times W)\frac{1}{T_s(t)}\frac{d(H - H_s(t))}{dt} = E\left(\frac{\rho_a^2 g}{C_P T_a}\right)^{1/3} \dot{Q}_c(t)^{1/3} H_s(t)^{5/3}. \tag{7.6}$$

The solution for the smoke layer height $H_s(t)$ thus requires an estimate of two more unknowns, the temperature of CV_1, $T_s(t)$ and the convective heat release rate $(\dot{Q}_c(t))$. $T_s(t)$ can be obtained by an energy balance in CV_1. Assuming that all the convective heat from the fire is delivered to the smoke the energy balance on CV_1 can be written as:

$$\dot{Q}_c(t) = \dot{m}_a C_P(T_s(t) - T_a). \tag{7.7}$$

Substituting for $\dot{m}_a$ using Eq. (7.3),

$$\dot{Q}_c(t) = E\left(\frac{\rho_a^2 g}{C_P T_a}\right)^{1/3} \dot{Q}_c(t)^{1/3} H_s(t)^{5/3} C_P(T_s(t) - T_a). \tag{7.8}$$

Equations (7.6) and (7.8) can now be solved numerically to obtain the descent of the smoke layer, $H_s(t)$, and its temperature, $T_s(t)$, as a function of time for the growing fire. The only missing piece is an expression for the convective heat release rate $\dot{Q}_c(t)$. For a steady fire source, $\dot{Q}_c(t)$ can be assumed to be a constant greatly simplifying the solution. Another approach is to use a growth rate expressed as a power law:

$$\dot{Q}_c(t) = \alpha t^n, \tag{7.9}$$

where α and n are constants that can be obtained from the fuel characteristics. Usually, classification is based on a slow, medium, fast, and ultra-fast growing fire where α increases and $n = 2$. Figure 7.4 shows typical values for α and n adapted from

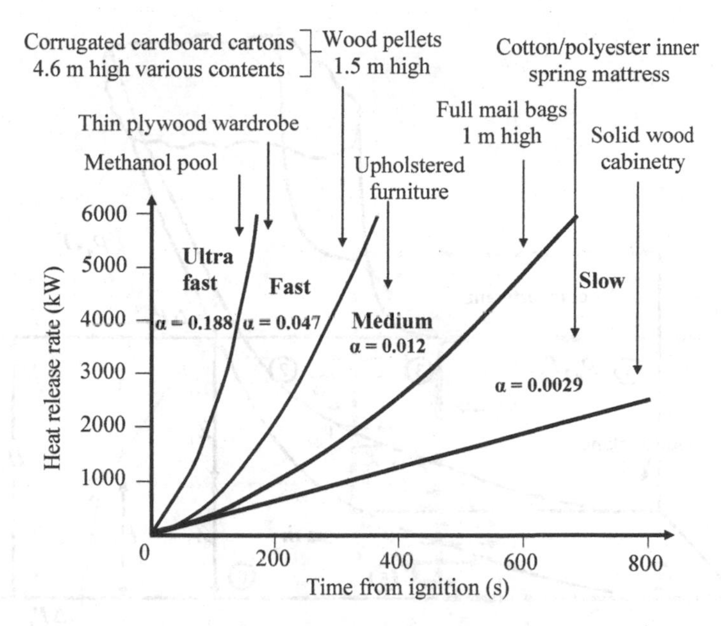

Fig. 7.4 Growth rates in a fire per standard codes based on a fitted power law given by $\dot{Q}_c(t) = \alpha t^n (\alpha \text{ in kJ/s}^3, n = 2)$

standard fire codes. For example, an enclosure with modern thin plywood wardrobe will have an ultra-fast growth rate as shown in Fig. 7.4. This is because thin plywood will sustain a fast flame spread thereby causing the fire to grow very quickly. On the other hand, an enclosure with a solid wood cabinetry will have a slow growth rate as shown in Fig. 7.4 because solid wood though combustible does not sustain a fast flame spread.

7.3 Flow of Gases In and Out of a Compartment

Figure 7.5 shows a schematic of the airflow in an enclosure that is in the fully developed stage (last photograph on Fig. 7.1). The fire-induced flow field is buoyancy dominated and turbulence causes efficient mixing of the products of combustion with the entrained air. It is assumed that the enclosure has a uniform gas temperature over its entire volume. In other words, the gases in the enclosure are well mixed. In actuality, the gas properties are stratified, and a two-zone model is more correct (this will also be discussed eventually). However, as a starting point, let us analyze the behavior of a fully developed fire first, where the upper zone (smoke layer) has increased to fill the compartment. The hot gases rise up and the smoke flows out from the top portion of the compartment. Figure 7.5 shows an enclosure whose temperature and density are T_g and ρ_g, and a vent of height H_0. The outside temperature and

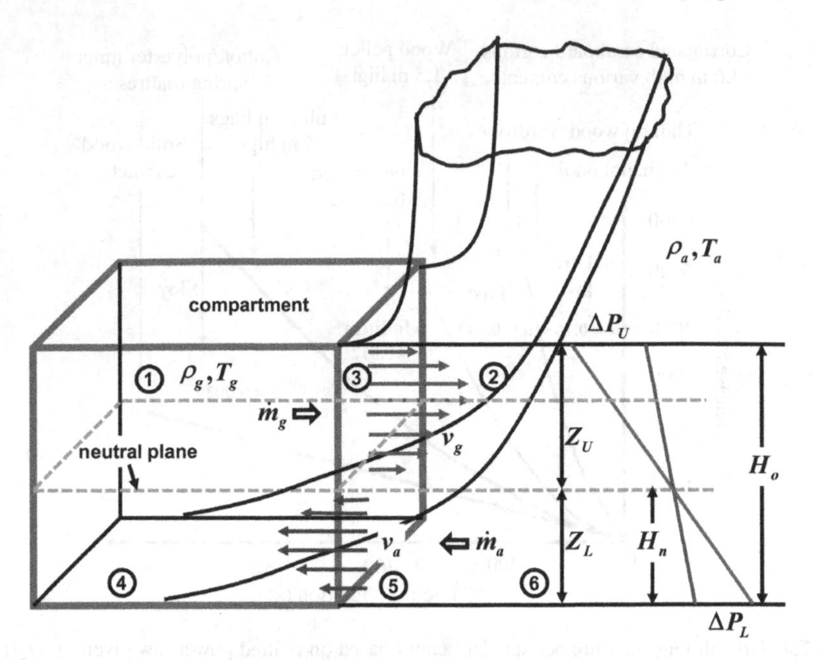

Fig. 7.5 Flow of gases in and out of a compartment during the fully developed stage

density are denoted by T_a and ρ_a. The hot gases flow out through the top part of the opening and the cold gases flow in through its lower part. The flow is caused because of the pressure differences that exist across the opening which are caused because of the temperature differences because of the fire. At some height above a reference height, the pressure difference will be zero and this height is called the neutral plane as shown in Fig. 7.5 and denoted by H_N. Hot gases will flow out above the neutral plane (shown as $\dot{m}_g$) and cold air will flow in below the neutral plane (shown as $\dot{m}_a$). To obtain mass flow out of the compartment, let us consider points 1, 2, and 3 as shown in Fig. 7.5. Also, let us assume that the flows are uniform across the width and velocity is only a function of height where the pressure changes because of buoyancy effects.

Applying Bernoulli's across 1 and 2 gives:

$$P_1 + \frac{1}{2}v_1^2\rho_g + Z_u\rho_g g = P_2 + \frac{1}{2}v_2^2\rho_a + Z_u\rho_a g, \tag{7.10}$$

where P_1 and P_2 are the pressure, v is the velocity, ρ_g is the density of the gases in the enclosure, and ρ_a is the ambient gas density. Now, the velocity of the gases deep inside the enclosure and far away into the ambient (v_1 and v_2) equals zero. Thus, Eq. (7.10) simplifies to

$$P_1 - P_2 = Z_u g(\rho_a - \rho_g), \tag{7.11}$$

or

$$\Delta P_U\Big|_{max} = Z_u g(\rho_a - \rho_g), \tag{7.12}$$

giving the maximum pressure difference in the upper vent. Similarly, applying Bernoulli's across 1 and 3 gives

$$P_1 + \frac{1}{2}v_1^2\rho_g + Z_u\rho_g g = P_3 + \frac{1}{2}v_3^2\rho_g + Z_u\rho_g g. \tag{7.13}$$

Again, noting that $v_1 = 0$, simplifies equation (7.13) to

$$P_1 - P_3 = \frac{1}{2}v_3^2\rho_g, \tag{7.14}$$

or

$$v_3 = \sqrt{\frac{2(P_1 - P_3)}{\rho_g}}. \tag{7.15}$$

Now, $P_1 - P_3 = \Delta P_U\Big|_{max} = Z_u g(\rho_a - \rho_g)$, thereby allowing us to express the maximum velocity through the upper vent in the form of density difference and given by,

$$v_3 = \sqrt{\frac{2Z_u g(\rho_a - \rho_g)}{\rho_g}}. \tag{7.16}$$

Similarly, by applying Bernoulli's equation across points 4, 5, and 6 gives v_5.

$$v_5 = \sqrt{\frac{2Z_L g(\rho_a - \rho_g)}{\rho_a}}, \tag{7.17}$$

v_5 is the maximum velocity through the lower vent of cold air flowing into the compartment. The velocity as a function of height above the neutral plane can be written as

$$v_g(z) = \sqrt{\frac{2Zg(\rho_a - \rho_g)}{\rho_g}}$$

and below the neutral plane equals

$$v_a(z) = \sqrt{\frac{2Zg(\rho_a - \rho_g)}{\rho_a}}.$$

These expressions can be integrated over the height to give the net velocity in the upper and lower regions of the compartment during a fully developed fire where the pressures have reached a steady state. It should be noted that the velocity is not constant across the vent area and changes with height. The mass flow rate out will occur across a height H_0 to H_N and given by

$$\dot{m}_{out} = C_d \int_0^{Z_u} \rho_g v_g(z) dA = C_d \int_0^{Z_u} \rho_g v_g(z) W dz, \tag{7.18}$$

where W is the width of the opening which is assumed to be a constant. We assume that the velocity is constant across the width of the vent. Note that we have defined $z = 0$ as the neutral plane. Hence, the limits of the integral range from the neutral plane ($z = 0$) to the top of the opening $z = Z_u$. Noting that ρ_g and W are constant, Eq. (7.18) can be simplified as

$$\dot{m}_{out} = C_d W \rho_g \int_0^{Z_u} v_g(z) dz, \tag{7.19}$$

and substituting $v_g(z)$ derived earlier gives,

$$\dot{m}_{out} = C_d W \rho_g \int_0^{Z_u} \sqrt{\frac{2Zg(\rho_a - \rho_g)}{\rho_g}} dz = C_d W \rho_g \sqrt{\frac{2g(\rho_a - \rho_g)}{\rho_g}} \int_0^{Z_u} \sqrt{z} dz. \tag{7.20}$$

Therefore,

$$\dot{m}_{out} = C_d W \rho_g \sqrt{\frac{2g(\rho_a - \rho_g)}{\rho_g}} \left(\frac{2}{3} Z_u^{3/2}\right). \tag{7.21}$$

Similarly, the mass flow into the compartment occurs in the region from H_N to the bottom of the opening expressed as Z_L in Fig. 7.2. Following the same procedure, the mass flow of cold air in the compartment can be shown to be equal to:

$$\dot{m}_{in} = C_d W \rho_a \sqrt{\frac{2g(\rho_a - \rho_g)}{\rho_a}} \left(\frac{2}{3} Z_L^{3/2}\right). \tag{7.22}$$

Equations 7.21 and 7.22 give the mass flow of hot gases out of the compartment and cold air into the compartment through the single opening in the enclosure. It should be noted that several simplifications were used in deriving the expressions, mainly steady-state, uniform temperature in the compartment, and uniform velocity across the width of the enclosure. It is also assumed that the mass of fuel vapor is negligible and a small fraction of $\dot{m}_{out}$. This is usually true for compartment fires where the mass of fuel is only 1–5% of $\dot{m}_{out}$ and can thus be ignored. The final unknown in the problem is the definition of reference height $z = 0$ or the height of the neutral plane itself. One equation linking the two unknown parameters Z_u and Z_L is

$$Z_u + Z_L = H_0. \tag{7.23}$$

The second expression can be obtained from conservation of mass given by

$$\dot{m}_{out} = \dot{m}_{in}, \tag{7.24}$$

or

$$C_d W \rho_a \sqrt{\frac{2g(\rho_a - \rho_g)}{\rho_g}} \left(\frac{2}{3} Z_u^{3/2}\right) = C_d W \rho_a \sqrt{\frac{2g(\rho_a - \rho_g)}{\rho_a}} \left(\frac{2}{3} Z_L^{3/2}\right), \tag{7.25}$$

which can be simplified as,

$$\frac{\rho_g}{\sqrt{\rho_g}} Z_u^{3/2} = \frac{\rho_a}{\sqrt{\rho_a}} Z_L^{3/2}, \tag{7.26}$$

or

$$\left(\frac{Z_u}{Z_L}\right)^{3/2} = \left(\frac{\rho_a}{\rho_g}\right)^{1/2}. \tag{7.27}$$

Substituting the expression for H_0 gives

$$\frac{H_0 - Z_L}{Z_L} = \left(\frac{\rho_a}{\rho_g}\right)^{1/3}, \tag{7.28}$$

which can be simplified to obtain an expression for Z_L given by,

$$Z_L = \frac{H_0}{1 + \left(\frac{\rho_a}{\rho_g}\right)^{1/3}}. \tag{7.29}$$

As Z_L is known, it can be substituted in Eq. (7.29) to obtain the mass flow of air into the compartment given by

$$\dot{m}_{in} = \frac{2}{3} C_d W \rho_a \sqrt{\frac{2g(\rho_a - \rho_g)}{\rho_a}} \left[\frac{H_0}{1 + (\rho_a/\rho_g)^{1/3}}\right]^{3/2}, \tag{7.30}$$

which can be simplified to

$$\dot{m}_{in} = \frac{2}{3} C_d A_0 \sqrt{H_0} \sqrt{2g} \rho_a \sqrt{\frac{(\rho_a - \rho_g)\rho_a}{\left[1 + (\rho_a/\rho_g)^{1/3}\right]^3}}, \tag{7.31}$$

where $A_0 = W \times H_0$ is the area of the vent. The parameter in the square root sign is called density factor and can also be expressed in the form of temperature using ideal gas law:

$$\sqrt{\frac{(\rho_a - \rho_g)\rho_a}{\left[1 + (\rho_a/\rho_g)^{1/3}\right]^3}} = \frac{353}{T_a}\sqrt{\frac{(T_a - T_g)}{\left[1 + (T_a/T_g)^{1/3}\right]^3}}. \tag{7.32}$$

The density factor reaches an asymptotic value of 0.214 as $T_g/T_a \rightarrow 2.72$. This is a useful result as during fully developed compartment fires, the gas temperature inside the enclosure usually exceeds 300 °C which is around the range when density factor reaches 0.214. Further, taking $C_d = 0.7$, $g = 9.81\ \text{m/s}^2$, $\rho_a = 1.2\ \text{kg/m}^3$, Eq. (7.31) simplifies to the classic equation

$$\dot{m}_{in} = \dot{m}_{out} = 0.5 A_0 \sqrt{H_0}. \tag{7.33}$$

The equation shows that the mass flow rate of air inside a compartment during the fully developed Regime I fire is only a function of the vent height and vent area.

7.4 The MQH Correlation

The general development of the classic Eq. (7.33) needs some discussion. The approach used to derive the equation was to compute the velocity from Bernoulli's equation. Then using ideal gas law, we substituted temperature for density and finally we performed an integration using the neutral plane as a reference to compute the mass flow rate. Rockett [8] has derived a similar expression for a developing (growth stage in Fig. 7.1) fire given by:

$$\dot{m}_{out} = \frac{2}{3} C_d A_0 \sqrt{H_0} \sqrt{2g} \rho_a \left[\theta(1-\theta)\right]^{1/2} \left[\frac{1 - H_n}{H_0}\right]^{3/2}, \tag{7.34}$$

and

$$\dot{m}_{in} = \frac{2}{3} C_d A_0 \sqrt{H_0} \sqrt{2g} \rho_a \left[1-\theta\right]^{1/2} \left(\frac{H_n}{H_0} - \frac{H_s}{H_0}\right)^{1/2} \left(\frac{H_n}{H_0} + \frac{H_s}{2H_0}\right), \tag{7.35}$$

where θ equals the ratio of ambient to compartment temperature (T_a/T_g). Compared to the simpler expression earlier, two additional parameters, H_n (the neutral plane height) and H_s (the smoke layer height), are needed to solve the mass in and out of the compartment. This is because for the developing/growing fire, the enclosure can no longer be considered one zone and a two-zone model is needed where the neutral plane height and smoke layer height need to be solved in order to compute the mass flow rates. The problem thus requires a numerical approach.

The ventilation factor has been further utilized by McCaffrey et al. [9] to develop the famous McCaffrey, Quintiere, and Harkleroad (MQH) correlation where using a simple heat balance applied to the layer of hot gases below the ceiling and the fact that the mass flow of gases in a fire is simply given by the ventilation factor. Using an energy balance, $\dot{Q}_c$, the convective heat release rate is given by,

$$\dot{Q}_c = (\dot{m}_{in} + \dot{m}_f)C_P(T_g - T_a) + h_k A_T(T_g - T_a), \tag{7.36}$$

where A_T is the total internal surface area of the enclosure (walls, ceiling, floor), neglecting openings and h_k is a global heat transfer coefficient that has been a subject of several experimental studies in compartment fire literature. For now, let us assume h_k is a constant and understand how the simple energy balance Eq. (7.36) allows us to express the two primary non-dimensional quantities controlling the problem. Noting that $\dot{m}_f \ll \dot{m}_{in}$, and can be neglected, the energy balance can be rearranged to give:

$$\frac{\Delta T}{T_a} = \frac{\dot{Q}_c/(C_P T_a \dot{m}_{in})}{1 + h_k A_T/(C_P \dot{m}_{in})}, \tag{7.37}$$

and using $\dot{m}_{in} \propto g^{1/2}\rho_a A_0 H_0^{1/2}$, $\Delta T/T_a$ becomes a function of two dimensionless groups given by

$$\frac{\Delta T}{T_a} = f\left(\frac{\dot{Q}_c}{(C_P\rho_a)T_a A_0\sqrt{gH_0}}, \frac{h_k A_T}{(C_P\rho_a)A_0\sqrt{gH_0}}\right), \tag{7.38}$$

or

$$\frac{\Delta T}{T_a} = C X_1^N X_2^M, \tag{7.39}$$

where X_1 and X_2 represent the two dimensionless groups in Eq. (7.39), and the constant C and exponents N and M are determined from experimental data. McCaffrey et al. [9] analyzed data from more than 100 experimental fires (from eight series of tests involving gas burners, wood cribs, plastic cribs, and liquid fuel fires) with size of the enclosures from conventional room fires to 1/8 size with varying construction material with a wide range of thermal properties. The constants C, N, and M were found so that Eq. (7.39) could be written as:

$$\frac{\Delta T}{T_a} = 1.63\left(\frac{\dot{Q}_c}{(C_P\rho_a)T_a A_0\sqrt{gH_0}}\right)^{2/3}\left(\frac{h_k A_T}{(C_P\rho_a)A_0\sqrt{gH_0}}\right)^{-1/3}. \tag{7.40}$$

Using $g = 9.81$ m/s^2, $\rho_a = 1.2$ kg/m^3, $T_a = 293$ K, and $C_P = 1.05$ kJ/kg-K, the expression can be simplified to give

$$\frac{\Delta T}{T_a} = 6.85\left(\frac{\dot{Q}_c^2}{h_k A_T A_0\sqrt{H_0}}\right)^{1/3}. \tag{7.41}$$

Note that since the expression is empirical, care must be taken to use the correct units, T is in K and $\dot{Q}_c$ is in kW, h_k is in kW/m^2-K, and all dimensions are in m. Further, the correlation is valid only for cases where upper gas layer temperatures

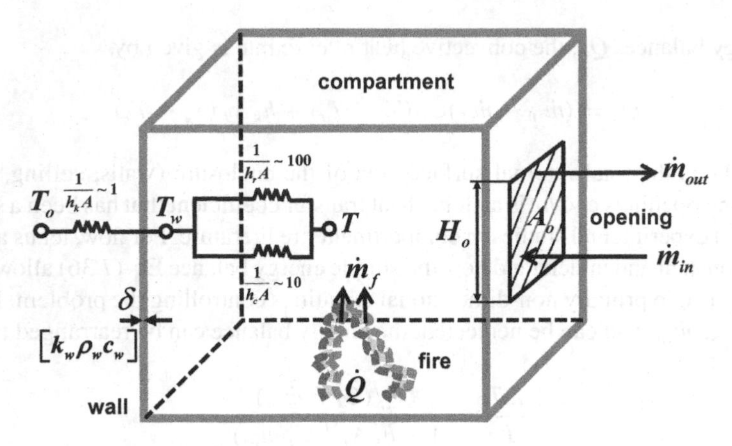

Fig. 7.6 Heat transfer through the wall in a compartment fire. In most cases, $T_w \sim T$

do not exceed 600 °C. Above 600 °C, flaming can occur intermittently within the smoke layer and many of the assumptions inherent in the model break down. As discussed in Chap. 6, when entrainment of cold air into the plume is restricted by a wall or a corner, the flame height increases and higher temperatures are observed in the upper smoke layer. Mowrer and Williamson [10] have shown that for fires flush with walls, Eq. (7.41) should be multiplied by 1.3 and for a corner fire by 1.7. While all properties are easy to determine, the estimation of h_k is involved as it depends on the duration of the fire and the thermal characteristics of the compartment boundary. As shown in Fig. 7.6, two situations can arise

1. For a fire which burns with a characteristic time (t_c) greater than the *thermal penetration time* (t_p) of the boundary, $t_p = \delta^2/4\alpha$, where α is the thermal diffusivity $(k/\rho C)$ and δ is the thickness of the wall, h_k is given as,

$$h_k = k/\delta, \tag{7.42}$$

   where k is the thermal conductivity of the material of the compartment wall.
2. If t_c is less than t_p, the boundary will be storing heat during the fire and little will be lost through the outer surface. Normally, this would require detailed solution of the transient heat conduction equations but a simplification can be achieved by replacing δ by $(\alpha t_c)^{1/2}$, the effective depth of the lining material which is heated significantly during the course of the fire also called as the thermal thickness and,

$$h_k = \frac{k}{(\alpha t_c)^{1/2}} = \left(\frac{k\rho C}{t_c}\right)^{1/2}. \tag{7.43}$$

For a compartment bounded by different lining materials, the overall value of h_k must be weighted according to the areas: thus, if the walls and ceiling are of a different material to the floor, then, if $t_c > t_p$:

$$h_k = \frac{A_{W,C}}{A_T}\frac{k_{W,C}}{\delta_T} + \frac{A_F}{A_T}\frac{k_F}{\delta_F}. \tag{7.44}$$

Here, δ_T is the thickness of the walls and ceiling (assumed equal) and δ_F is the thickness of the floor. $A_{W,C}$ is the area of the walls and ceiling, A_F is the area of the floor, and A_T is the total surface area of the compartment minus area of the vents. But if $t_c < t_p$,

$$h_k = \frac{A_{W,C}}{A_T}\left(\frac{(k\rho C)_{W,C}}{t_c}\right)^{1/2} + \frac{A_F}{A_T}\left(\frac{(k\rho C)_F}{t_c}\right)^{1/2}, \tag{7.45}$$

where $(k\rho C)_{W,C}$ is the thermal inertia of the wall and ceiling, and $(k\rho C)_F$ is the thermal inertia of the floor material.

7.5 Flashover

The MQH correlation can also be used to predict whether a fire in a given enclosure may cause flashover. If a temperature rise of 500 °C is taken as a conservative criterion for the upper layer gas temperature at the onset of flashover then substitution for X_1 and X_2 in Eq. (7.40) gives:

$$\dot{Q}_c = \left[\sqrt{g}(C_P\rho_a)T_a^2\left(\frac{\Delta T}{480}\right)^3\right]^{1/2}\left(h_k A_T A_O\sqrt{H_O}\right)^{1/2}. \tag{7.46}$$

With $\Delta T = 500$ K, $\dot{Q}_{FO} = 610(h_k A_T A_O\sqrt{H_O})^{1/2}$, where $\dot{Q}_{FO}$ (kW) is the rate of heat output necessary to produce a hot layer at approximately 500 °C beneath the ceiling. The square root dependence indicates that if there is 100% increase in any of the parameters h_k, A_T, or A_w, then the fire will have to increase in heat output by only 40% to achieve the flashover criterion as defined. For highly insulating linings, such as fiber insulating board, or expanded polystyrene, the size of the fire sufficient to produce flashover is greatly reduced, even if the linings are assumed to be inert and any contribution to the rate of heat release is neglected.

7.6 Fire Severity—Average Smoke Temperature

An equation that gives the average temperature of the smoke exiting the compartment is obtained from Eq. (7.41) as given by

$$\Delta T = C_T \left[\frac{\dot{Q}_c^{\,2}}{(hA)A_0\sqrt{H_0}} \right]^{1/3}, \tag{7.47}$$

where $\Delta T = T - T_a$, $C_T = 6.85$ for centered fires, 12.4 for corner fire, and 16.5 for room linings, $\dot{Q}_c$ = heat release rate in kW, h = heat loss coefficient kW/m^2°C, A = interior surface area in m^2, A_0 = area of vent in m^2, and H_0 = height of vent in m.

The heat loss coefficient, h, is taken as the maximum of $\sqrt{\frac{k\rho C}{t}}$ or $\frac{k}{l}$, where $k\rho C$ is the interior construction thermal inertia (kW/m^2°C)^2-s, k is the thermal conductivity, l is the interior construction thickness, and t is the time (duration of the fire). The solution to the equation requires a method to estimate $\dot{Q}$, hA, and $A_0\sqrt{H_0}$ also called as a ventilation factor.

7.7 Compartments with Ceiling Vents

Unlike studies with wall vents in compartments which has been extensively studied in literature, fire dynamics of compartments with ceiling vents is an open area of fire research with few papers in this interesting area [11, 12]. The main difference between compartments with vertical wall ventilation and compartments with horizontal ceiling ventilation is the nature of the fire-induced vent flows at the openings and the corresponding impact on the heat release rate and compartment temperatures. For fires in vertically vented enclosures (wall vents such as doors and windows), a stratified two-way flow exists in which the smoke flows out of the upper part of the vent and fresh air flows inside from the lower part. However, when a single vent is located at the ceiling interesting flow patterns emerge as shown in Fig. 7.7.

Very few researchers have investigated this pattern in detail. In general, when the vent is at the ceiling three conditions are typically observed [11]:

1. Choked followed by extinguishment,
2. Erratic pulsing also sometimes referred to as ghosting flames, and
3. Strong steady burning.

Emmons [12] and Cooper [13] analyzed the airflow through the ceiling vent using the concept of a flooding pressure given by

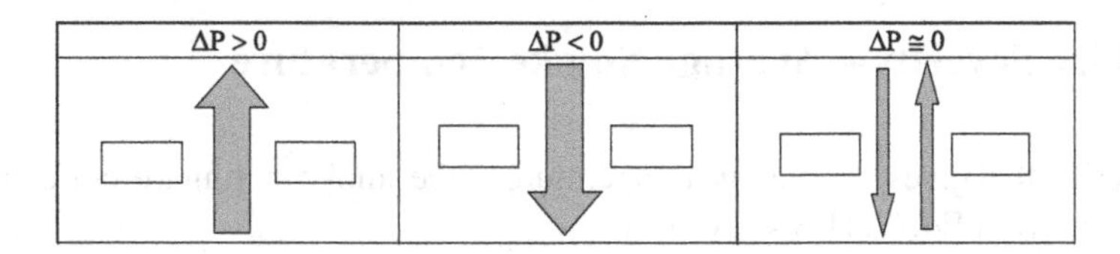

Fig. 7.7 Flow direction due to pressure difference for ceiling vent [11], where ΔP is the pressure difference between inside and outside of the compartment

$$|\Delta P_f| = 0.0873\frac{\Delta\rho^2 gD}{C_d^2\rho_f}, \tag{7.48}$$

where

$$D = 4A_v/L$$

A_v is the vent area, and L equals the perimeter of the ceiling vent. $g = 9.8$ m/s^2 and ρ_f is the density of air calculated at the compartment temperature. The flow coefficient (C_d) for flow through a ceiling vent is given by:

$$C_d = 0.138 + 1.12 \times 10^{-4} Re. \tag{7.49}$$

The expression for flooding pressure can then be used to obtain the flow in and out of a compartment with a ceiling vent.

7.8 Case Study

Consider a triangular opening as shown below for vent in a compartment of height H_0, and base width W.

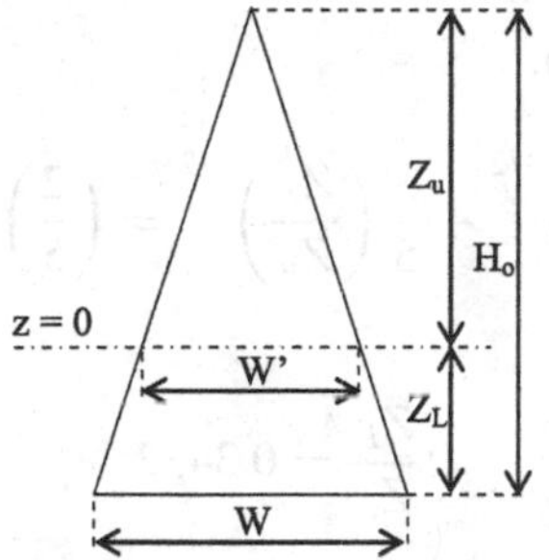

The location of the neutral plane from the base can be evaluated as follows. Applying Bernoulli's equation,

$$v_g(z) = \sqrt{\frac{2z(\rho_a - \rho_g)g}{\rho_g}}.$$

Using Eq. (7.18),

$$\dot{m}_{out} = C_d \int_0^{Z_u} \rho_g v_g(z) W(z) dz$$

$$\dot{m}_{out} = C_d \rho_g \int_0^{Z_u} \sqrt{\frac{2z(\rho_a - \rho_g)g}{\rho_g}} W' \left(1 - \frac{z}{Z_u}\right) dz$$

$$\dot{m}_{out} = C_d \rho_g W' \sqrt{\frac{2(\rho_a - \rho_g)g}{\rho_g}} \left(\frac{2}{3} Z_u^{1.5}\right) \left(\frac{2}{5}\right).$$

Similarly,

$$v_a(z) = \sqrt{\frac{2z(\rho_a - \rho_g)g}{\rho_a}}.$$

Using Eq. (7.18)

$$\dot{m}_{in} = C_d \int_0^{Z_L} \rho_a v_a(z) W(z) dz$$

$$= C_d \rho_a \int_0^{Z_L} \sqrt{\frac{2z(\rho_a - \rho_g)g}{\rho_a}} \left[W' + (W - W') \left(\frac{z}{Z_L}\right)\right] dz$$

$$\dot{m}_{in} = C_d \rho_a W' \sqrt{\frac{2(\rho_a - \rho_g)g}{\rho_a}} \left(\frac{2}{3} Z_L^{1.5}\right) \left[1 + \frac{3}{5} \left(\frac{W}{W'} - 1\right)\right].$$

Since $\dot{m}_{out} = \dot{m}_{in}$, we obtain,

$$\left(\frac{Z_L}{Z_u}\right)^{3/2} + \frac{3}{5} \left(\frac{Z_L}{Z_u}\right)^{5/2} = \left(\frac{2}{5}\right) \sqrt{\frac{\rho_g}{\rho_a}}.$$

As $T_g/T_a \to 2.72$

$$\frac{Z_L}{Z_u} = 0.3432$$

$$\frac{Z_L}{H_0 - Z_L} = 0.3432 \Rightarrow Z_L = 0.2555 H_0.$$

We can compare this value to the location of neutral axis for a rectangular opening, which is evaluated using Eq. (7.29) and is found to be $0.4174 H_0$. The case study shows how the vent flow equations can be applied to vents with different geometries.

7.9 Conclusion

Analysis of a fire in a building is one of the core topics in fire. This chapter demonstrated that the primary controlling parameter for the evolution of the heat release rate in a compartment fire is the ventilation, or more specifically $A_0\sqrt{H_0}$ where A_0 and

H_0 are the area and height of the ventilation opening. When the ventilation opening is small which is true for most compartments, the fire is ventilation controlled and the compartment temperature at steady state is a function of the air flow in and out of the compartment which $\sim 0.5A_0\sqrt{H_0}$ kg/s. The simple expression is very useful in assessing the fire boundary condition for structural fire safety. When the ventilation opening is very large, such that there is plenty of air availability compared to quantity of fuel, the fire is fuel limited. These types of fires are an active area of fire research.

Review Questions

1. Describe the different stages in fire development. Consider a polynomial fit to the temperature evolution from initial to the fully developed stage as

$$T = T_0 + C_1 t + C_2 t^2 + C_3 t^3,$$

where T_0 is the initial temperature and C_1, C_2, and C_3 are constants. Identify the value of t at which flash over will occur and the value of t where a fully developed fire will occur.
2. What is the difference between the two regimes in compartment fires?
3. Two materials A and B burn in a compartment and a power law fit is done to obtain the convective heat release rate given as

$$HRR = \alpha t^2.$$

Determine the value of α_A and α_B such that they have the same HRR of 1000 kW at time $t_A = 200$ s and $t_B = 600$ s. Which material (A or B) is more hazardous in terms of fire growth rate?
4. What is the MQH correlation and its applications? Also mention limitations of the correlation.
5. What is ventilation factor and how is the average smoke temperature dependent on it?
6. What are the conditions observed based on the pressure difference for a ceiling vent?

References

1. Kawagoe, K., and T. Sekine. 1963. Estimation of Fire Temperature Rise Curves in Concrete Buildings (Part 1, 2). *Transactions of the Architectural Institute of Japan* 85: 38–43.
2. Harmathy, T. 1972. A new look at compartment fires Part I. *Fire Technology* 8 (3): 196–217.
3. Thomas, P. 1965. The rate of temperature rise in a compartment fire. *Fire Safety Science* 595: 1–1.

4. Emmons, H. W. 1979. The prediction of fires in buildings, *Proceeding of Combustion Institute*, vol. 17(1), 1101–1111. Elsevier.
5. Karlsson, B., and J.G. Quintiere, 1999. *Enclosure fire dynamics*. CRC Press.
6. Torero, J.L., A.H. Majdalani, C. Abecassis-Empis, and A. Cowlard. 2014. Revisiting the compartment fire. *Fire Safety Science* 11: 28–45.
7. Thomas, P.H., A. Heselden, and M. Law. 1967. Fully-developed Compartment Fires: Two Kinds of Behaviour: HM Stationery Office.
8. Rockett, J.A. 1976. Fire induced gas flow in an enclosure. *Combustion Science and Technology* 12 (4–6): 165–175.
9. McCaffrey, B., J. Quintiere, and M. Harkleroad. 1981. Estimating room temperatures and the likelihood of flashover using fire test data correlations. *Fire Technology* 17 (2): 98–119.
10. Mowrer, F.W., and R.B. Williamson. 1987. Estimating room temperatures from fires along walls and in corners. *Fire Technology* 23 (2): 133–145.
11. Wakatsuki, K. 2001. *Low ventilation small-scale compartment fire phenomena: Ceiling vents, College Park*. MD: University of Maryland.
12. Emmons, H. 1996. A universal orifice flow formula. *Thirteenth meeting of the UNJR panel of fire research and safety*.
13. Cooper, L. Y., Calculation of the flow through a horizontal ceiling/floor vent. 1989. *National institute of standards and technology*. MD: Gaithersburg.

Chapter 8
Dust Flames

Many materials that are virtually non-flammable in bulk form become highly explosive if dispersed like a cloud of fine particles in the air. In industries that manufacture, transport, process, and use combustible dust, accidental dust deflagrations represent a hazard to both personnel and equipment. "Fire" caused by micron-sized particles (dust) causes extensive material damage, injury, and loss of life in coal mines, wood, paper, chemical, food, grain, and pharmaceutical plants. The current codes and standards used in dust explosion protection design are based on empirical correlations developed 50 years ago. More alarmingly, they are mainly based on theory developed for gas explosions. In short, the difference between a *gas* and a *dust* fire is not fundamentally understood. The purpose of this chapter is to explain the current status of our understanding of a dust flame and provide an insight to the student on the several controlling parameters associated with this complex problem. A complete model for a dust flame does not exist, but the fundamental building blocks are reasonably understood, as shown in this chapter.

A particle is usually called dust if its diameter is less than 500 μm. Three types of dust, coal, sand, and sodium bicarbonate, are shown in Fig. 8.1 as small piles comprising particles in a size range of 75–90 μm. Figure 8.1d shows the same coal dust suspended in the air using a 1 cm diameter nozzle. Coal dust is combustible, sand dust is inert, and sodium bicarbonate releases CO_2 when heated by the flame, thereby serving as an active suppressant. Many organic dusts, like wheat flour, rice flour, paper, wood, sugar, etc., are highly flammable when dispersed in the air as a fine cloud of particles. In fact, the first documented explosion and fire occurred in 1785 in a bakery store room where an oil lamp ignited the flour particles suspended in the air after a bag was accidentally dropped on the floor and burst into a cloud of dust. Table 8.1 shows some of the recent dust explosion incidents from across the world. The ever-advancing and expanding chemical, metallurgical, mining, and pharmaceutical industries have given birth to a steadily increasing number of new finely divided flammable materials. The accidents occurring because of dust explo-

A. S. Rangwala and V. Raghavan, *Mechanism of Fires*,
https://doi.org/10.1007/978-3-030-75498-3_8

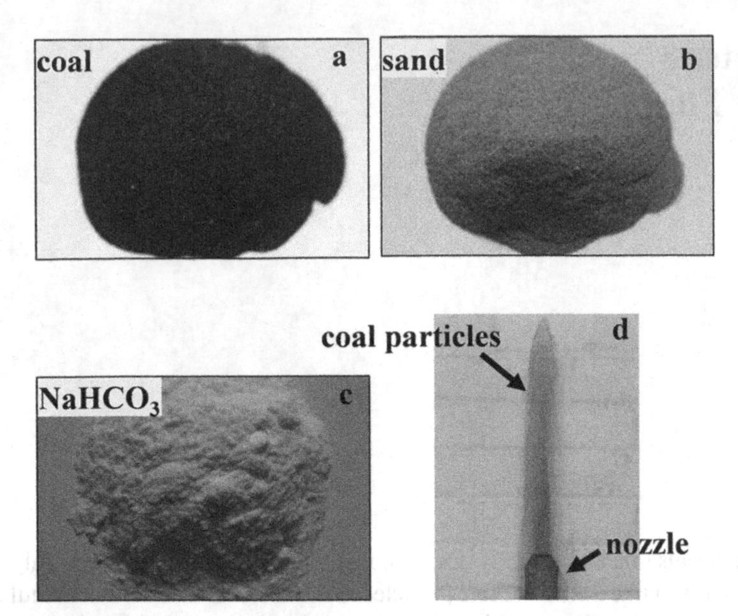

Fig. 8.1 Three types of dusts encountered in flames: **a** Reactive such as coal, **b** inert like sand, and **c** reactive but not combustible such as sodium bicarbonate. In **d**, the coal dust is suspended using a nozzle. The mass flow rate of the dust particles shown equals 0.002 g/sec. In each photo, the particle size is in the range of 75–90 μm

sions show a lack of fundamental knowledge of particle-air combustion in a turbulent environment, forming an active research area in fire safety.

Figure 8.2a shows a sketch of a dust deflagration with the phenomena controlling the movement of the flame through an unburned mixture of dust and oxidizer. Four areas of interest labeled 1–4 in Fig. 8.2a are depicted as (1) dust lifting and entrainment, (2) the dynamics of the dust cloud, (3) burning of the dust, and finally (4) the influence of confinement which coupled with reaction zone producing high temperature and rapid expansion of gases can cause development of high pressures. Most structures collapse when pressures rise above atmospheric by only 0.1 atm (1.5 psi), while closed vessel dust explosions can generate pressures up to 8–9 atm (120 psi)!

Figure 8.2b and c, adapted from Sun et al. [1], shows two mechanisms that control the propagation of the reaction zone, where the flame can propagate as a sequential ignition of diffusion or partially premixed flames (like a flame jumping from one fuel element to another like in an array) or as a premixed flame purely in the gas phase. Both the mathematical theory and the parameters controlling these two modes of propagation are completely different!

In practice, a dust flame will propagate in between these regimes. This adds a further level of complexity to the study of dust explosions. This is because the pressure rise is related to the rate of energy production, which is related to the flame propagation rate. Due to these complications, a comprehensive mathematical theory

Table 8.1 Recent explosion incidents involving dusts

Industry	Fuel	Location	Year	Deaths
Coal mine	Methane and coal dust	Cucunuba, Colombia	2017	11
Coal mine	Methane and coal dust	Badong, China	2016	11
Coal mine	Methane and coal dust	Chifeng, China	2016	32
Chemical company	Cement production	Qinghai Province, China	2016	6
Wood flour mill	Wood dust	Cheshire, UK	2015	4
Color play	Starch dust	New Taipei, Taiwan	2015	15
Car parts	Metal dust	Jiangsu Province, China	2014	75
Coal mine	Methane and coal dust	Donetsk, Ukraine	2013	7
Saw mill	Wood dust	Lakeland Mills saw mill in Prince George, Canada	2012	2
Saw mill	Wood dust	Babine Forest Products in Burns Lake, Canada	2012	2
Coal mine	Methane and coal dust	Sardinata, Colombia	2011	21
Powder manufacturer	Iron dust	Hoeganaes Corporation, TN, USA	2011	5
Coal mine	Methane and coal dust	Upper Branch mine, West Virginia, USA	2010	29
Coal mine	Methane and coal dust	Pike River, New Zealand	2010	29

to predict explosion mechanisms of dust clouds is at present beyond reach and forms an interesting research topic for future students.

With the complexities associated with the basic structure of a dust flame, the problem of quantifying a dust explosion hazard therefore reduces to exploring the parameters controlling each of the four zones shown in Fig. 8.2a, the inter-relationships between them, and researching experimental methods to obtain them. In this chapter, we will begin by exploring the structure of a dust flame, understanding the thermophysical aspects controlling the burning behavior, and then end with the current methodologies used by industry to evaluate the dust hazard. The student will appreciate the fundamental problem and its complexities, and ultimately be able to understand how current engineering codes and standards rely on large factors of safety to circumvent the lack of fundamental property data related to dust-air flames.

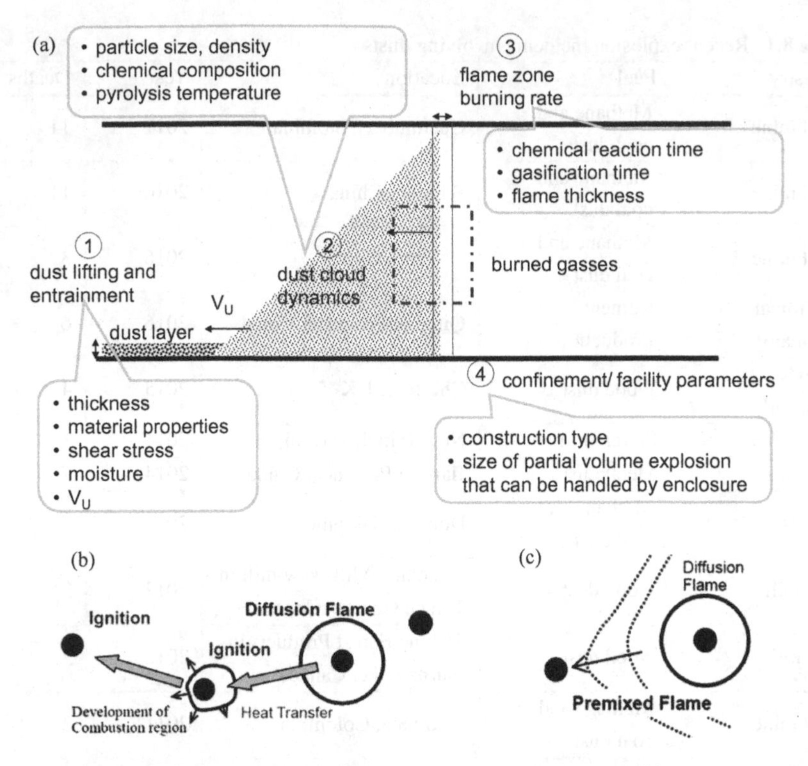

Fig. 8.2 Schematic of **a** a dust flame showing four controlling zones of interest, two extreme cases of flame propagation mechanisms possible in the flame/burning region (zone 3) where the dust-oxidizer flame can move through **b** sequential ignitions, or **c** as a premixed flame

8.1 Flame Structure—What is a Dust Flame?

The earliest work on dust-air premixed flames is provided by Nusselt [2] and the most up-to-date review on the topic is given by Eckhoff [3]. Smoot and Horton [4], Krazinski et al. [5], and Slezak et al. [6] have investigated the problem using numerical modeling. A thorough analysis of the flame structure of a laminar and turbulent dust flame is discussed in Rockwell and Rangwala [7, 8], Xie et al. [9], Lee et al. [10], and Ranganathan et al. [11, 12]. Unlike a premixed gas flame, a mixture of dust and oxidizer involves a multiphase flow, which causes difficulty in both experiments and modeling. Radiation heat transfer also plays a significant role in most cases and is important in dust flames.

For gases, since the reactants are separated by molecular distances, premixed combustion is guaranteed to small scales. In contrast, dust flames involve the combustion of a dust-air suspension. A dust cloud, which is uniform when viewed at a macro-scale (e.g., cloud radius), may not be premixed at a small scale (e.g., inter-

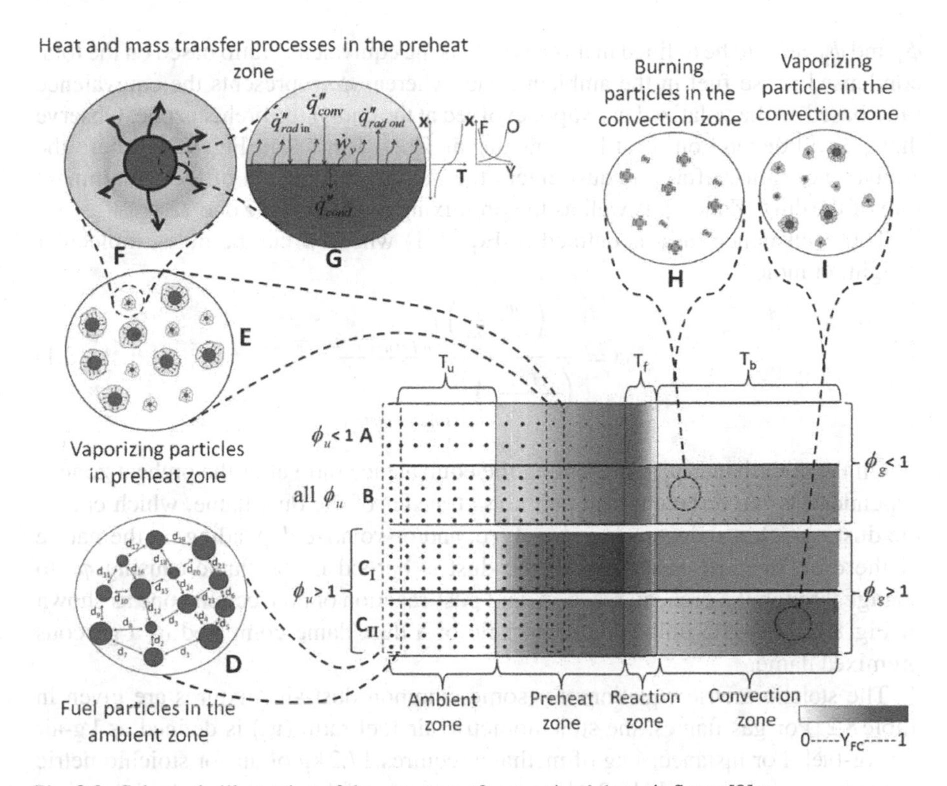

Fig. 8.3 Schematic illustration of the structure of a premixed dust-air flame [8]

particle distance). This caused researchers to make a distinction between two types of dust flames [3]: the Nusselt flame and the volatile flame. In the Nusselt flame, strictly heterogeneous combustion occurs at the surface of the particles, sustained by diffusing oxygen towards the particles' surface. Therefore, a Nusselt flame, which on a macroscopic scale may seem like premixed combustion, consists of an ensemble of local diffusion flames.

Such flames are not considered in the theoretical models for dust flame propagation [8, 14] and no theory exists currently. With a volatile flame, particles produce vapor before combustion. As shown in Fig. 8.3, when mixed with air, these gases and vapors burn as a premixed gas in three distinctive mechanisms:

1. Devolatilization and burning of volatiles followed by combustion of a solid residue.
2. Melting followed by evaporation and subsequently vapor phase burning.
3. Evaporation through a solid oxide shell followed by combustion of the vapor outside the shell.

Figure 8.3 shows four potential scenarios (A, B, C_I, and C_{II}) that are possible given the three mechanisms discussed earlier. Because of particles, two equivalence ratios,

ϕ_u and ϕ_g, need to be defined in this case. ϕ_u is the equivalence ratio based on the total condensed-phase fuel in the ambient zone, whereas ϕ_g represents the equivalence ratio based on the volatized gas vapor evolved at the end of the preheat zone. Observe that ϕ_u will depend on what happens during Zone 1 and 2 in Fig. 8.2a, where the equivalence ratio before the dust enters the flame is a function of the entrainment rate of the dust (Zone 1) as well as the premixing dynamics in Zone 2.

The equivalence ratio is defined in Eq. (8.1) where m can be mass, molecular weight, or moles.

$$\phi = \frac{\left(\frac{m_{fuel}}{m_{oxidizer}}\right)\Big|_{mixture}}{\left(\frac{m_{fuel}}{m_{oxidizer}}\right)\Big|_{stoichiometric}}. \tag{8.1}$$

On the other hand, ϕ_g representing the equivalence ratio after the preheat zone is dependent on the turbulent heat and mass transfer of the dust flame, which causes the dust particles to decompose, pyrolyze, and/or volatize depending on the nature of the dust. In most cases, not all the dust is burned in the flame causing ϕ_g to change through the preheat, reaction, and post-reaction or convection zone as shown in Fig. 8.3. This is a unique characteristic of a dust flame compared to a gaseous premixed flame.

The stoichiometric equations for some common dust-air mixtures are given in Table 8.2. For gas flames, the stoichiometric air-fuel ratio (ν_g) is defined as kg-air per kg-fuel. For instance, 1 kg of methane requires 17.2 kg of air for stoichiometric combustion, and ν_g equals 17.2. However, in the case of dust reactions, the theoretical concentration defined as mass (in grams) of dust per unit volume of surrounding air forms a key parameter for determining the stoichiometric condition. Typical values for theoretical concentration for different dust-air mixtures are shown in Table 8.2. For comparison with methane gas, coal dust which contains 40% methane has a theoretical concentration of 173 g/m^3.

An example how the theoretical concentrations in Table 8.2 were derived are shown below for PMMA:

PMMA:

$$C_5H_8O_2 + 6(O_2 + 3.76N_2) \rightarrow 5CO_2 + 4H_2O + 22.56N_2$$

$$MW_{PMMA} = 100 \text{ kg/kmol}$$

$$\left(\frac{F}{A}\right)_{st} = \frac{m_F}{m_A} = \frac{1 \times 100}{6 \times (32 + 3.76 \times 28)} = 0.1214$$

$$\phi = \frac{(F/A)}{(F/A)_{st}} = \frac{1 \times \rho_F/\rho_A}{0.1214},$$

where "1" in the numerator denotes that 100% of the particle will be converted into fuel vapor. With coal dust considered in the next example this value is 0.4 as only

Table 8.2 Stoichiometric chemical reaction and theoretical stoichiometric dust concentration for some mixtures

Dust	Equation	Theoretical concentration (g/m^3)
Aluminium	$2Al + 1.5(O_2 + 3.76N_2) \rightarrow Al_2O_3 + 5.64N_2$	310
Silicon	$Si + (O_2 + 3.76N_2) \rightarrow SiO_2 + 3.76N_2$	241
Iron	$2Fe + 1.5(O_2 + 3.76N_2) \rightarrow Fe_2O_3 + 5.64N_2$	150
Niacin Amide	$C_6H_6N_2O + 7(O_2 + 3.76N_2) \rightarrow 6CO_2 + 3H_2O + 26.32N_2$	150
Lycopodium Spores (1.1% ash and 3.5% natural water)	$CH_{1.68172}O_{0.21941}N_{0.01664}S_{0.00044} + 1.311165(O_2 + 3.76N_2) \rightarrow CO_2 + 0.84086H_2O + 0.00044SO_2 + 4.9383N_2$	120
Glucose	$C_6H_{12}O_6 + 6(O_2 + 3.76N_2) \rightarrow 6CO_2 + 6H_2O + 22.56N_2$	259
Coal (40% CH_4, 60% Ash)	$CH_4 + 2(O_2 + 3.76N_2) \rightarrow CO_2 + 2H_2O + 7.52N_2$	173
Coal (40% CH_4, 60% C)	$CH_4 + 2C + 4(O_2 + 3.76N_2) \rightarrow 3CO_2 + 2H_2O + 15.04N_2$	86
PMMA	$C_5H_8O_2 + 6(O_2 + 3.76N_2) \rightarrow 5CO_2 + 4H_2O + 22.56N_2$	144
Cellulose	$C_6H_{10}O_5 + 6(O_2 + 3.76N_2) \rightarrow 6CO_2 + 5H_2O + 22.56N_2$	233
Methane gas	$CH_4 + 2(O_2 + 3.76N_2) \rightarrow CO_2 + 2H_2O + 7.52N_2$	69

40% of the particle volatilizes to fuel vapor. Coming back to PMMA, when $\phi = 1$, $\rho_F = 0.1214\rho_A$

$$\rho_A = \frac{p}{R_A T} = \frac{101325}{287.058 \times 298} = 1.184 \text{ kg/m}^3.$$

Theoretical concentration, to achieve $\phi_u = 1$

$$\rho_F = 1.184 \times 0.1214 = 0.144 \text{ kg/m}^3 = 144 \text{ g/m}^3.$$

The uniqueness of a dust flame is the definition of ϕ.

In the unburned side, $\phi_u = 1$ means that all the particles in ambient zone when volatilized will give $\rho_F = 0.1214\rho_A = 0.1214 \times 1.184$ kg/m^3 $= 0.144$ kg/m^3 or 144 g/m^3. If more than 144 g/m^3 is available on the unburned side then $\phi_u > 1$ and much more fuel is theoretically available. However, because the fuel needs to pyrolyze and only the fuel vapor concentration is important, depending on the rate of pyrolysis $\dot{\omega}_v'''$, ϕ_g can take any value up to a maximum equal to ϕ_u. If $\phi_g < \phi_u$ which is usually the case as not all particles will volatilize in the short residence time available in the preheat zone, the dust particles will also be present in the convection zone, where burned products of combustion exist at high temperature. These particles will continue to volatilize but because of lack of oxygen will not react. Further, radiation losses will play an important role because of high-temperature particles in the convection zone. These aspects are not common in gas-phase premixed flames, and unique to dust flames only.

For coal dust [Coal (40% CH_4, 60% Ash)]:

$$CH_4 + 2(O_2 + 3.76N_2) \rightarrow CO_2 + 2H_2O + 7.52N_2$$

$$MW_{CH_4} = 16 \text{ kg/kmol}$$

$$\left(\frac{F}{A}\right)_{st} = \frac{m_F}{m_A} = \frac{1 \times 16}{2 \times (32 + 3.76 \times 28)} = 0.0583$$

$$\phi = \frac{(F/A)}{(F/A)_{st}} = \frac{0.4 \times \rho_F/\rho_A}{0.0583} (40\% \text{ fuel})$$

when $\phi = 1$, $\rho_F = 0.0583\rho_A/0.4$

$$\rho_A = \frac{p}{R_A T} = \frac{101325}{287.058 \times 298} = 1.184 \text{ kg/m}^3.$$

Theoretical concentration, to achieve $\phi_u = 1$

$$\rho_F = 1.184 \times 0.0583/0.4 = 0.173 \text{ kg/m}^3 = 173 \text{ g/m}^3.$$

In Fig. 8.3, label A denotes the condition where $\phi_u < 1$ and $\phi_g < 1$. Label B denotes the condition where ϕ_u can take any value but with the condition of $\phi_g < 1$. It is assumed here that the number of particles is large enough such that the flame standoff distance surrounding each particle is significantly larger than the characteristic separation distance between the particles. For liquid droplets (spray flames), this is observed for overall equivalence ratios (ϕ_u) greater than 0.7 [15]. Label C denotes the condition where both ϕ_g and ϕ_u are greater than 1. Two cases can arise, in this case: C_I, where the particles completely vaporize in the preheat zone (burned mixture in convective zone comprises fuel-rich vapor), and C_{II} where the particles do not completely vaporize and act as heat sinks in the convective zone.

Insets D to E are used to further explain the structure of the three mechanisms shown in Fig. 8.3. The inset labeled D shows a close-up of the ambient zone where the random distribution of both particle separation and size in a dust-air mixture is highlighted. In all dust flame models, these quantities are approximated as average values, and it is assumed that there is an even distribution of particles of an average size and a single fuel type. The implications of this simplification have not been thoroughly investigated in the combustion literature, though there has been some research into deriving models for dual particle sizes [16] and dual particle fuels [17].

The inset labeled E shows a close-up of the preheat zone. The differences in particle size play a significant role as smaller particles heat up faster and vaporize almost completely, while larger particles continue to be in the condensed phase as they move into the reaction and convection zones. The inset labeled F shows a close-up view of a single vaporizing particle with the inset labeled G showing the surface of a particle in the preheat zone where the fuel changes phase from solid to gas and premixes with the oxidizer to establish a flame front. At this stage, it is possible that burning is localized on the surface alone. Note that the change in phase (solid to liquid to gas) slows down burning velocity significantly compared with a gas flame.

Further, as shown in G (Fig. 8.3), the heat flux related to in-depth conduction ($\dot{q}''_{cond}$) and radiation ($\dot{q}''_{rad,in}$, $\dot{q}''_{rad,out}$) also plays a significant role. The inset G also shows the vaporization rate ($\dot{w}'''_v$), which is determined by an energy balance of the net heat transfer divided by heat of gasification. These additional parameters influence the burning of particle-air flames as discussed further in Fig. 8.4, which shows a sketch of the flame structure for the four types of equivalence ratio combinations considered in A, B, C_I, and C_{II}.

The profiles of mass fraction of condense-phase fuel (Y_s), mass fraction of vaporized fuel (Y_{FC}), the vaporization rate ($\dot{\omega}'''_v$), the reaction rate ($\dot{\omega}'''_F$), and the temperature (T), across ambient, preheat, reaction, and convection zones, are shown in Fig. 8.4. Case A represents the conditions where $\phi_g < 1$ and all the condensed-phase fuel vaporizes as shown in Fig. 8.3 (A). When $\phi_g < 1$, fuel is the limiting reactant and is consumed in the reaction zone. Vaporization mainly takes place in the preheat zone, with the mass fraction of the condense-phase particles (Y_s) dropping to zero and the mass fraction of the fuel vapor (Y_{FC}) reaching a maximum in the preheat zone. The temperature increases through the preheat zone, reaches the maximum value in the reaction zone, and remains constant in the convection zone (losses are neglected).

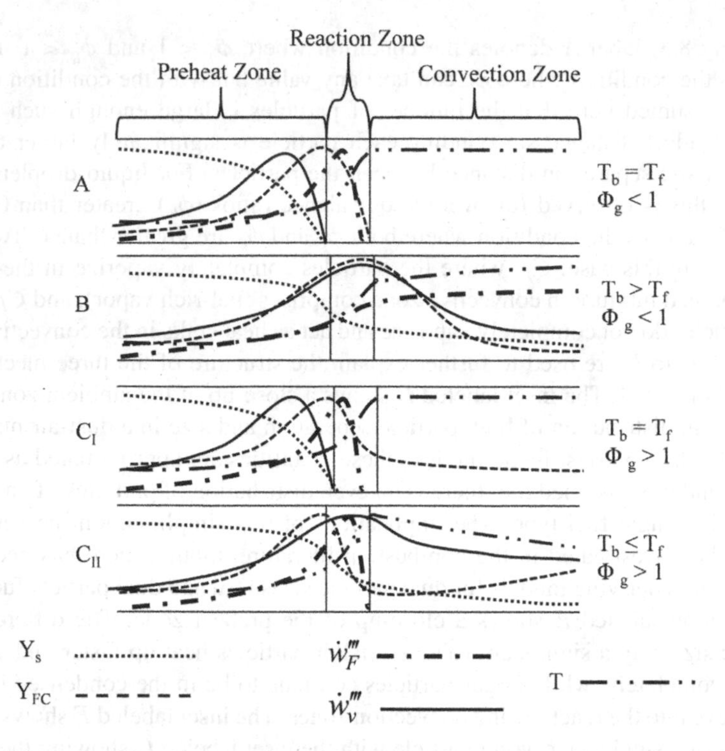

Fig. 8.4 Schematic of flame structure in a dust-air flame [8]

Case B represents the conditions of $\phi_g \leq 1$ and particles do not vaporize in the preheat zone as shown in Fig. 8.3b. Particles continue to burn even in the convection zone, resulting in the temperature in the convective zone (T_b) to be greater than the flame temperature (T_f). The inset labeled H in Fig. 8.3 shows the convection zone in Case B, where the fuel particles continue to burn in the presence of excess oxygen. The presence of these particles is mainly because of their larger sizes and/or slow vaporization rate. Note that this case results in increase of the temperature in the convection zone.

Case I represents the condition where $\phi_g \geq 1$. All the particles vaporize in the preheat zone as shown in Fig. 8.3 (C_I). However, in this case, only part of the gas-phase fuel burns in the flame zone, because of the fuel richness of the mixture, and there is fuel vapor left over in the convection zone. Oxygen is the limiting reactant in this case. Assuming there are no heat losses, temperature can be approximated to remain constant in the convection zone.

Case II represents the conditions where $\phi_g \geq 1$ and the particles are not completely vaporized in the preheat zone as shown in Fig. 8.3 (C_{II}). Similar to Case I, oxygen is the limiting reactant for this case also. However, as the particles continue to vaporize in the convective zone, the mass fraction of the fuel vapor increases and the temperature in the convection zone decreases. The inset labeled I in Fig. 8.3

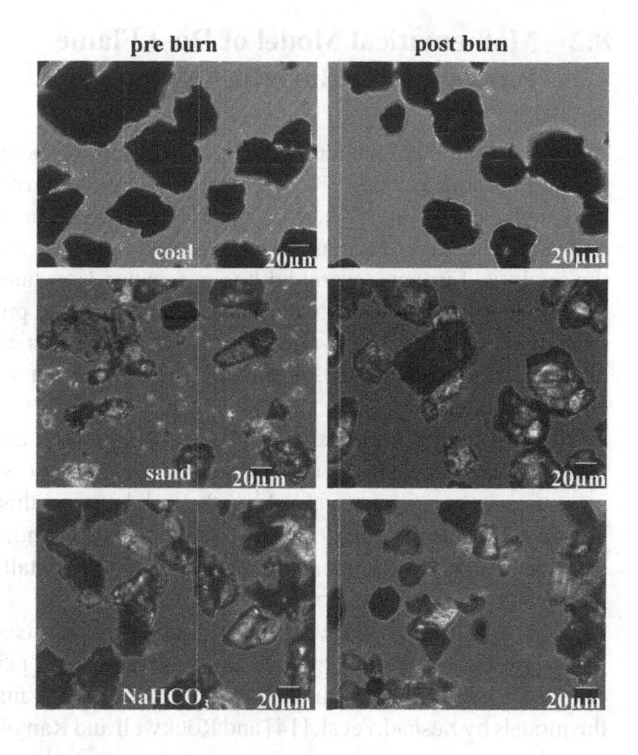

Fig. 8.5 Dust pre-burn and post-burn microscopic images [14]

shows a close-up of the convection zone, which occurs in Case II where fuel particles continue to vaporize but do not burn because of oxygen limitation. This continued vaporization increases the fuel vapor mass fraction and decreases the convection zone temperature.

Figure 8.5 shows the pre-burn and post-burn photographs of dust particles during laminar premixed combustion [18].

The particles were collected from a Bunsen burner type of premixed burner where 40–70 μ dust particles were injected in a lean methane-air premixed flame. It is clear that the particles are not consumed in the flame because of the short residence times.

In the Pittsburg seam coal (bituminous coal) particle case, for example, the sharp edges of the particles are gone and particles become more rounded. It is also observed that particles show small craters, which are sites for gas diffusion. A similar behavior is observed in sodium bicarbonate particles as well, where the sharp corners become more rounded. Sand particles behave as an inert and therefore no significant change in size is observed. The equivalence ratio for most practical dust-air flames is usually much greater than 1. In other words, most dust explosions are fuel rich and B and C_{II} (Figs. 8.3 and 8.4) are realistic cases. The other flame structures A and C_I can be created in laboratory flames by externally controlling the preheat temperatures as discussed in Lee et al. [10].

8.2 Mathematical Model of Dust Flame Propagation—Burning Velocity

The fundamental parameter that controls the damage potential of an explosion is the laminar burning velocity. Table 8.3 shows a summary of the expressions developed by researchers in the past. It is clear that there is no universal theory and each analysis is limited to the assumptions used and/or experimental conditions during the analysis. Further, several ad hoc parameters/constants are used to make theory match experimental data. Note that unlike a gaseous premixed flame, the laminar burning velocity of a dust-air flame is defined by particle entrainment (particulate mechanics, aerodynamics), gasification (heat and mass transfer), and combustion characteristics.

Lack of fundamental experiments makes the laminar burning velocity for dust flames a parameter poorly understood and current assessment methodologies leave an enormous gap while establishing the link between this fundamental variable and design criteria for ventilation and seals as well as methodologies for prevention and damage control. The student can look through the details of each derivation in the relevant papers cited in Table 8.3.

Table 8.3 shows that the burning velocity of a premixed particle-air flame is controlled by several parameters such as vaporization rate, particle size, number density, etc. To explore the controlling parameters, a sensitivity analysis was performed using the models by Seshadri et al. [14] and Rockwell and Rangwala [8]. The reason the two models were chosen is because they represent both lean and rich local equivalence ratios (ϕ_g) thereby covering all the cases A to C_{II} described earlier. The sensitivity analysis is performed using Saltelli et al. [34].

A sensitivity parameter (S_i) is defined as follows:

$$S_i = \frac{ln(y_{1,i}) - ln(y_{1,0})}{ln(x_{1,i}) - ln(x_{1,0})},$$

where x_0 and x_1 are the initial and final values of the parameter to be examined, and y_0 and y_1 are the corresponding model solutions in burning velocity. The normalized sensitivity analysis value (S_{iN}) is calculated by dividing the calculated S_i value by $S_{i,max}$, representing the largest value calculated for the model. The normalized sensitivity is used to distinguish the controlling parameters of the fuel lean (scenario A or B shown in Figs. 8.3 and 8.4) and fuel-rich models (C_I and C_{II} scenarios shown in Figs. 8.3 and 8.4).

The model representing each and listed in Table 8.3 is tested using a constant set of parameters listed in Table 8.4. Each parameter alters by 5% and the resulting sensitivity is shown in Fig. 8.6.

Table 8.3 Laminar burning velocity expressions for particle-air flames

Reference	Expression for S_L	Variables and assumptions
Seshadri et al. [14]	$(v_u)^2 = \frac{2(1+b)v_f \lambda_u C_6 C_8^2}{\rho_u C_7} exp\left(-\frac{E}{RT_f}\right)$ $b = \frac{Y_F/Y_{FC}}{C_8}, Y_{FC}Q = C_7(T_F - T_u)$ $C_7 = C_p + \frac{4\pi r^3 C_s \rho_s n_s}{3\rho}, C_8 = \frac{RT_f^2}{E(T_f - T_u)}$	*Assumptions*: Negligible heat of vaporization, $\phi_g < 1$, *Variables*: S_L—burning velocity, v_f—velocity of fuel b—scaled mass fraction of fuel, λ_u—Thermal conductivity (unburned) ρ—density of ambient reaction stream C_6—frequency factor C_7—heat capacity of mixture C_8—expansion parameter E—activation energy of the reaction R—gas constant, T_f—maximum temperature Y—mass fraction, Q—heat release rate T_u—Ambient temperature Ze—Zeldovich number C_p—heat capacity of gas, r—fuel particle radius C_s—heat capacity of fuel particle ρ_s—density of fuel particle n_s—local number density of particles *Subscripts*: F—gaseous fuel, f—conditions at the reaction zone, u—ambient condition (unburnt)
Seshadri et al. [14]	$S_L = v_u exp\left(\frac{-qZe}{2}\right), q = Q_v/Q$	*Assumptions*: Small values of q *Variables:* v_u—burning velocity calculated neglecting heat of vaporization of fuel particles Q_v—heat associated with vaporizing unit mass of fuel
Rockwell et al. [8]	$S_L^2 = \frac{2\lambda_u B(Ze)^{-2}}{\rho_u C_{P,mix}} exp\left(-\frac{E}{RT_b}\right)$ $Y_{FC}Q = C_{P,mix}(T_f - T_u)$ $3a\alpha^{2/3} - 3a^2\alpha^{1/3} + a^3 - 1 = 0$	*Assumptions*: $\phi_g \geq 1$
Cashdollar [19]	$S_L = \frac{K_{st}}{4.84\left(\frac{P_{max}}{P_0} - 1\right) P_{max}}$	–

Table 8.3 (continued)

Reference	Expression for S_L	Variables and assumptions
Ballal [20]	$S_D = \frac{\alpha \Delta T_r}{\delta_r \Delta T_{pr}}, \alpha = \frac{k}{\rho C_P}$ $\delta_r = \alpha_g^{0.5}\left[[(I) + (II)]^{-1} - (III)\right]^{-0.5}$ $(I) = \frac{C_3^3 \rho_f D_{32}^2}{8 f^2 C_1 (k/C_P)_g \phi ln(1+B)}$ $(II) = \frac{\alpha}{S_L^2}\frac{\Delta T_r}{\Delta T_{pr}}, (III) = \frac{9qC_1^2}{C_{p,g}\rho_f C_3^3}\frac{\epsilon\sigma}{f D_{32}}\frac{T_p^4}{\Delta T_r}$ $C_1 = D_{20}/D_{32}, C_3 = D_{30}/D_{32}$	*Variables:* S_D—Burning Velocity of dust cloud α—thermal diffusivity ΔT_r—Temperature rise in reaction zone ΔT_{pr}—Temperature rise in preheat zone δ_r—flame thickness in reaction zone α_g—thermal diffusivity of gas (I)—diffusion term (II)—chemical kinetic term (III)—radiation heat loss term D_{20}—Surface area mean diameter D_{30}—volume mean diameter D_{32}—Sauter mean diameter ρ_f—density of the fuel f—swelling factor, k—thermal conductivity C_P—specific heat ϕ—equivalence ratio B—mass transfer number S_L—Laminar burning velocity α—thermal diffusivity q—fuel-air mass ratio ϵ—emissivity of fuel particles σ—Stefan-Boltzmann constant (5.66×10^{-8} W/m^{2}K^4)
Cassel [21]	$S_L = \frac{\mu \frac{T_b - T_i}{b} + \frac{b\omega\epsilon C_9 F\left(T_b^4 - T_u^4\right)}{\rho_d r}}{(C_P\rho + C_d\omega)(T_i - T_u)}$	*Assumptions*: Mallard and Le Chatelier's picture of combustion wave amended by an additional term to account for the effects of radiation *Variables*: μ—thermal conductivity T_u—temperature of the unburned T_b—temperature of the burned T_i—ignition temperature ϵ—emissivity of the particle surfaces C_9—radiation correction factor larger than 1 F—geometrical view factor b—burning zone thickness C_P—specific heat of the gas ρ—density C_d—specific heat of the dust ρ_d—density of the dust ω—concentration of dust r—average particle radius
Cassel [21]	$S_L^2 = \frac{K\rho_b}{\tau\rho_u}\frac{T_b - T_i}{T_i - T_u - \frac{\tau\omega\epsilon\alpha F\rho_u\left(T_b^4 - T_u^4\right)}{\rho_d\rho_b r(C_P\rho + C_d\omega)}}$ $K = \frac{\mu}{C_P\rho + C_d\omega}$	Introduces burning time of a single particle, includes the velocity of the burned mass relative to the flame front and the ratio of densities of the unburned and burned gases.

Table 8.3 (continued)

Reference	Expression for S_L	Variables and assumptions
Cassel et al. [21]	$\tau = \frac{\rho_d r^2 R(T_u)^{3/2}}{2MDp(T_a)^{1/2}},\ K = \frac{\rho_d \rho_u R T_u^{3/2}}{2M\rho_b T_a^{1/2}}$ $S_L^2 = \frac{KDp}{r^2 k} \frac{T_b - T_i}{T_i - T_u - \frac{r\omega k F\epsilon\alpha\left(T_b^4 - T_u^4\right)}{\rho_d Dp\left(C_P \rho + C_d \omega\right)}}$	*Assumptions*: Oxygen diffusion governs the burning of individual particles *Variables*: D—the diffusion coefficient at temperature T_u R—Gas Constant T_a—average ambient gas temperature p—the average oxygen partial pressure M—the oxygen equivalent of the fuel
Proust et al. [22]	$S_L = \frac{\lambda_g \frac{T_{ad} - T_{inf}}{\delta} + 2\delta\rho_P F\sigma_0 \frac{T_{ad}^4 - T_0^4}{\rho_d d_P}}{\rho_m C_m \left(T_{ad} - T_{inf}\right)}$ $\delta = S_L \tau_c$	*Assumptions*: Must include radiation because of view factor F *Variables*: λ_g—thermal conductivity of the gas phase T_{inf}—typical ignition temperature δ—flame thickness, τ_c—induction time d_p—particle diameter, F—view factor ρ_p—specific mass of dust T_{ad}—adiabatic flame temperature T_0—the initial temperature of the mixture ρ_m—specific mass of the medium C_m—specific heat of the medium σ_0—Stefan-Boltzmann constant
Dahoe et al. [23]	$S_{uL} = \left[1 + \frac{L}{R}\right] S_{uL}^0$	*Assumptions*: Weakly strained flames *Variables*: S_{uL}—laminar burning velocity L—Markstein length R—Radius of curvature of the flame sheet S_{uL}^0—Laminar burning velocity (unstretched)
Dahoe [13]	$S_L = \left[-2\frac{\lambda}{\rho_u C_P} \frac{\Delta H_c}{C_P\left(T_f - T_u\right)} \frac{\overline{\dot{w}_F}}{\rho_u}\right]^{1/2}$	*Assumptions*: Energy flux for heating of the unburned mixture to the flame temperature is controlled by the conduction of heat through the preheat zone. *Variables*: λ—thermal conductivity ρ_u—density of unburned fuel ΔH_c—heat of combustion of the fuel C_p—specific heat $\overline{\dot{w}_F}$—average fuel consumption rate, T_f—temperature of the burned T_u—temperature of the unburned

Table 8.3 (continued)

Reference	Expression for S_L	Variables and assumptions
Ballal [20]	$S_L = \frac{\delta_r}{t_c},\ t_c = t_q - t_e$ $t_e = \frac{C_3^3 \rho_f D_{32}^2}{8C_1 (k/C_P)_g ln(1+B)}$ $t_q = \left[\frac{\alpha_g}{\delta_r^2} + \left(\frac{9q}{C_{p,g}\rho_f}\right)\frac{C_1^2}{C_3^3}(f_1 D_{32})^{-1}\epsilon\sigma\frac{T_P^4}{\Delta T_r}\right]^{-1}$ $S_L = \frac{k_g}{C_{p,g}\rho_u \delta_r}\frac{\Delta T_r}{\Delta T_{pr}}$	*Assumptions*: Truly evaporating particles, not carbon or coal *Variables*: B—mass transfer number $C_1 = D_{20}/D_{32}$, $C_3 = D_{30}/D_{32}$ T_p—the particle temperature T_{pr}—the preheat zone temperature T_r—reaction zone temperature S_u—laminar burning velocity q—dust concentration k—the thermal conductivity δ_r—the thickness of reaction zone C_p—specific heat of gas ρ_u—density of unburned gas ρ_f—the density of particle f_1—the swelling factor ϵ—the emissivity of fuel particles α—the thermal diffusivity
Ogle et al. [24]	$S_L = \frac{1}{\rho_0 C_P (T_f - T_0)}\left[\sigma T_f^4 - \frac{3}{2}ka(T_f - T_0)\right]$	Radiation Conduction Convection Model
Ogle et al. [24]	$S_L = \frac{\sigma T_f^4}{\rho_0 C_P (T_f - T_0)}$	Radiation and convection model
Nomura and Tanaka [25]	$S_L = \frac{L}{\Delta T_\infty},\ L = \left[\frac{\pi}{6}D_P^3\frac{\rho_P}{C_d}\right]^{1/3}$	*Variables*: C_d—dust concentration D_p—particle diameter ρ_p—particle density
Hertzberg et al. [26, 27]	$S_L = \left(\frac{\alpha}{\tau}\right)^{1/2}$ $\alpha = \frac{q - \epsilon\sigma T_P^4 - \frac{1}{3r^2}\frac{\partial}{\partial t}(r^3 \rho_p h)}{T_P - T_g}$ $\tau = \tau_{dv} + \tau_{mx} + \tau_{pm}$	–
Nagy and Verakis [28, 29]	$S_L = S_{L,r}\left(\frac{T_u}{T_r}\right)^2\left(\frac{P_r}{P}\right)^\beta$	(1) Closed Vessel, (2) Ideal gas, (3) Point ignition is at the spheres center by a negligible energy supply, (4) Viscosity and heat capacity are constant, (5) Burning velocity is low compared to the velocity of sound, (6) Pressure is spatial uniform throughout vessel at any instant, (7) Thickness of propagating reaction zone is negligible compared to vessel radius, $S_{L,r}$—reference velocity at 300 K, $\beta <= 0.5$
Nomura and Tanaka [30]	$S_L = \frac{L}{\Delta T_\infty},\ L = \left[\frac{\rho_p}{C_d}\right]^{1/3} D_p$	*Assumption*: Closed Vessel ΔT_∞—a complex function of particle and combustion properties

Table 8.3 (continued)

Reference	Expression for S_L	Variables and assumptions
Ogle et al. [31]	$S_L = B + (B^2 + A)^{1/2},\ A = \frac{\lambda}{\rho_u C_P \tau}\left(\frac{T_f - T_i}{T_i - T_0}\right)$ $B = \frac{1}{2}\left(\frac{\epsilon\sigma}{C_P}\right)\left(\frac{T_f^4 - T_0^4}{T_i - T_0}\right)$	*Assumption*: Closed Vessel based on classical Mallard-Le Chatelier model [32] for premixed gases, with and additional term for thermal radiation *Variables*: ρ—initial density of the gas phase ϵ—emissivity T_i—ignition temperature T_0—initial temperature λ—thermal conductivity τ—the burning time
Goroshin et al. [33]	$\frac{3}{2}\mu\phi\frac{\xi}{\xi + K_+} = \frac{1}{\kappa^4}(\kappa_+ - \kappa_-)(\kappa_+ - Z)(\kappa^2 - Z)$ $Z = \frac{\kappa^2 Le}{2} - \sqrt{\left(\frac{\kappa^2 Le}{2}\right)^2 + \frac{3\kappa^2\phi Le}{2}}$ $K_\pm = \frac{(1+\nu\phi)\kappa^2}{2} \pm \sqrt{\left(\frac{(1+\nu\phi)\kappa^2}{2}\right)^2 + \kappa^2\eta}$	Fuel lean mixture model
Goroshin et al. [33]	$\frac{1}{\mu\phi} = \frac{\xi}{K_1(\epsilon + K_1)}\frac{1 - \exp(-K_1)}{\sqrt{1 + 4\frac{\eta}{\kappa^2}}}$ $K_1 = \frac{\kappa^2}{2}\left(1 + \sqrt{1 + 4\frac{\eta}{\kappa^2}}\right)$ $\kappa^2 = \frac{S_L^2\tau_c}{\lambda_{g,u}},\ \eta = \frac{2\bar{h}Nu\tau_c}{d^2},\ \mu = \frac{\omega_{st}Q}{\rho_{gu}C_g(T_{st} - T_u)}$ $\xi = \frac{3r_{g,u}\lambda_{g,u}\tau_c}{r_u^2\rho_s},\ \phi = \frac{\omega}{\omega_{st}}$	*Assumption*: Fuel lean mixture model *Variables*: λ_g—thermal conductivity of gas ω—dust concentration C—molar oxygen concentration D—mass diffusivity of oxygen d—diameter of the quenching channel E—Overall activation energy K_0—reaction rate pre-exponential factor Le—Lewis number, m—mass of particle Q—heat of reaction per unit fuel mass Nu—Nusselt number, r—particle radius t—time, S_L—burning velocity W_F—heat source term α—heat transfer coefficient (gas and wall) Δ—characteristic flame interval δ—overall flame thickness ϕ—equivalence ratio η—heat loss parameters κ—non-dimensional burning velocity $\bar{h}$—heat transfer coefficient $\nu = C_s B_{st}/c_g\rho_g$ τ_s—combustion time of individual separated particle at initial oxygen concentration

Table 8.4 Default parameter values for sensitivity analysis

$\phi = 1$	$A = 3.4 \times 10^{-6}$ g/(cm^{2}s-K)	$B = 3.5 \times 10^7$ (mol-s)$^{-1}$
$n = 1.33$	$\lambda_u = 3.5 \times 10^{-4}$ cal/(cm-s-K)	$\lambda_s = 9.5 \times 10^{-5}$ cal/(cm-s-K)
$q = 0.05$	$\rho_s = 1$ g/cm^3	$\rho_g = \rho_u$ g/cm^3
$\nu_F = 1, \nu_O = 2$	$Q = 1.55 \times 10^{-4}$ cal/g	$W_F = 16$ g/mol
$C_{p,s} = 1.357$ cal/(g-K)	$\rho_u = 1.77 \times 10^3$ g/cm^3	$R = 1.994$ cal/(K-mol)
$C_{p,g} = 0.24$ cal/(g-K)	$E = 23 \times 10^3$ cal/mol	$W_O = 32$ g/mol
$T_u = 300$ K	$r_s = 10^{-5}$ m	

Default values used to make Fig. 8.6 are shown in Table 8.4. Results show that in the scenarios representing $\phi_g < 1$ as shown in Fig. 8.3, (cases A and B), the exponent of the vaporization kinetic term (n) and convection zone temperature (T_b) are the most dominant in determining the burning velocity. Thus, when $\phi_g < 1$ the vaporization rate of the dust particle is controlling the rate of the overall combustion. In scenarios C_I and C_{II}, as shown in Figs. 8.3 and 8.4, the activation energy of the gas-phase reaction (E) is the most dominant followed by the frequency factor (B) and the heat of the reaction (Q). This is because of the fact that when $\phi_g > 1$ the system is oxygen controlled, and the terms related to the rate of the gas-phase reaction are the controlling parameters.

The burning velocity is not sensitive to the equivalence ratio based on the total condensed-phase fuel in the ambient zone, in the oxygen limited case. The particle size or particle radius is slightly more dominant for the fuel lean case. On the other hand, the thermal conductivity of the dust is more influential in the case of an oxygen-controlled flame front. The influence of heat of gasification is incorporated using $S_{L,q} = S_L \exp -qZ_e/2$, where $S_{L,q}$ is the burning velocity calculated including the heat of gasification (c.f. Seshadri et al. [14] for derivation) and q is the ratio of heat of gasification to heat of reaction (assumed to be small). It is observed that q may have some influence on the burning velocity in the lean case, but is not significant in the rich case.

8.3 The Explosion Sphere

Traditionally, the reactivity of explosive gas-air mixtures, flammable vapors, mists, and dust clouds has been characterized by the maximum rate of pressure rise also called deflagration index (K_{st}) and maximum pressure (P_{max}) determined in constant volume explosion vessels as shown in Fig. 8.7. These vessels are approximately spherical with 20 L to 1000 L capacity.

The concept of using an explosion sphere for flammability studies with premixed flames is based on the early experimental work by Andrews et al. [36], Abdel-Gayed

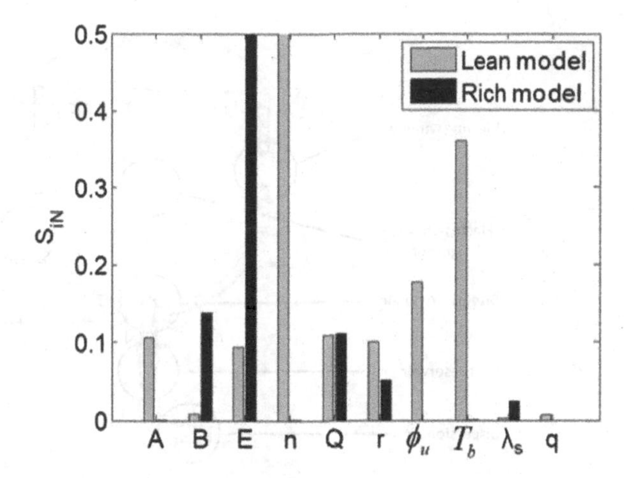

Fig. 8.6 Controlling parameters for dust-air premixed flame combustion. A sensitivity analysis showing the sensitivity of burning velocity of a particle-air flame using the model equations developed in Seshadri et al. [14] and Rockwell and Rangwala [8] (see Table 8.3)

and Bradley [37] and many subsequent publications by the Leeds group (cf. Bradley, [38]) where burning velocities of gas flames were measured by initiating a flame from a central spark and recording the spherical flame propagation in a closed vessel optically (via a quartz window) or by recording the pressure-time trace by placing a pressure transducer at the vessel walls as shown in Fig. 8.8. This setup was adopted for dust-air premixtures as discussed below (a detailed history is given in Eckhoff [39]).

The standard dust explosion vessel is equipped with a vacuum pump, dust dispersion, ignition, and pressure sensor systems. The test procedure begins by placing a measured quantity of a dust sample in a reservoir as shown in Fig. 8.7. Prior to ignition the dust-air mixture is discharged into the vessel through a fast acting valve and a rebound nozzle. The dispersed dust cloud is ignited after a specified ignition delay time. The ignition source is typically two chemical igniters, 5 kJ each, positioned near the center of the vessel. The main operating conditions for two typical explosion vessels (0.02 and 1 m^3) are shown in Table 8.5.

Measures for the energy content and the reactivity of the dust-air suspension are derived from the pressure-time history as shown in Fig. 8.8. Both the maximum pressure and deflagration index are determined from the same type of experiments in constant volume explosion vessels. Further details are described in standards: ISO 6184-1 [40], EN 14034-1 [41], EN 14034-2 [42], and ASTM E-1226 [43].

The explosion mitigating measures, such as explosion vents, suppression systems, etc., rely on empirical correlations involving parameters called deflagration indices (K_g for gas and K_{st} for dusts) and maximum pressure, P_{max}. The deflagration index is determined by an explosion sphere apparatus based on the maximum rate of pressure rise and the volume of the explosion sphere given by the cube root law:

$$K_{st} = \left(\frac{dP(t)}{dt}\right)_{max} (V_0)^{1/3},$$

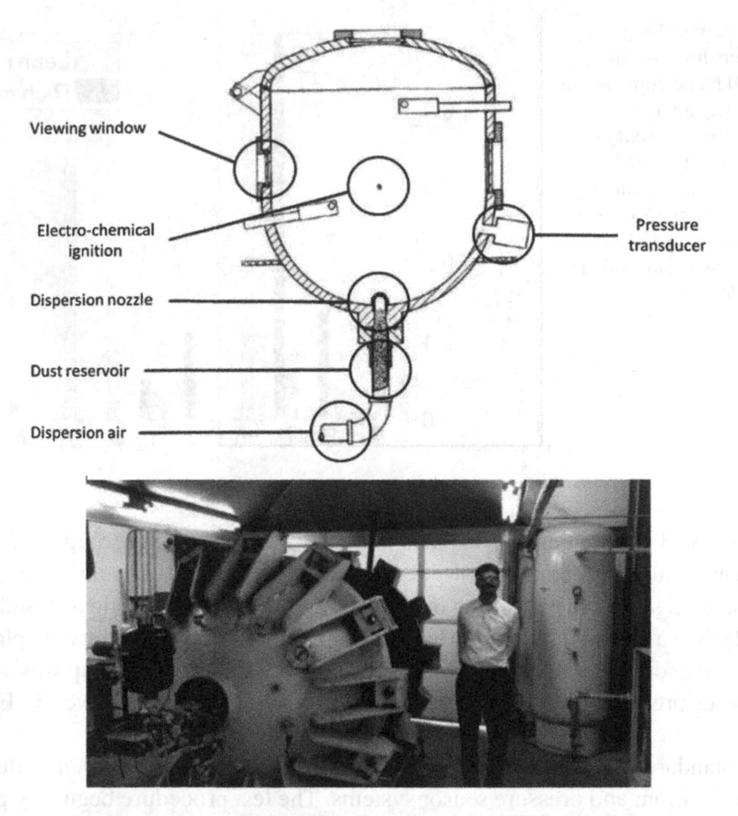

Fig. 8.7 Diagram of a 20 *L* explosion sphere (Image from ASTM E1515 [35]) and one of the authors standing next to a 1 m^3 explosion sphere at Fike Corporation, Bluesprings, MO, USA

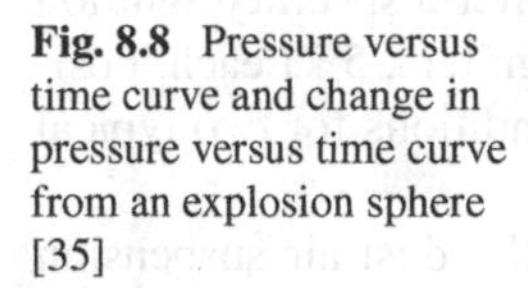

Fig. 8.8 Pressure versus time curve and change in pressure versus time curve from an explosion sphere [35]

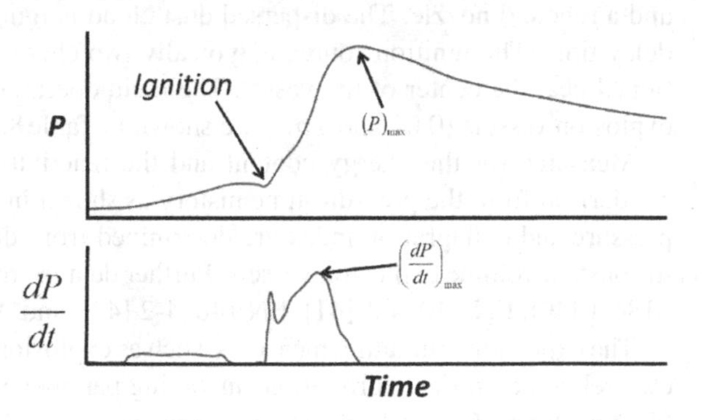

where V_0 is the volume of the explosion sphere and dP/dt is the change in pressure over time. It has been shown [3] that the deflagration index changes as the size of the explosion sphere changes. This makes the investigation of a dust flame rather difficult and also complicates the hazard classification as the quantity used to characterize

Table 8.5 Main operating conditions for two typical explosion vessels

Components	20 L vessel	36 L vessel
Reservoir volume	0.6 L	1 L
Initial pressure in vessel	−0.6 barg	−0.3 barg
Fast acting valve time	45 ms	50 ms
Pressure at time of ignition	0 barg	0 barg
Ignition delay time	60 ms	75 ms

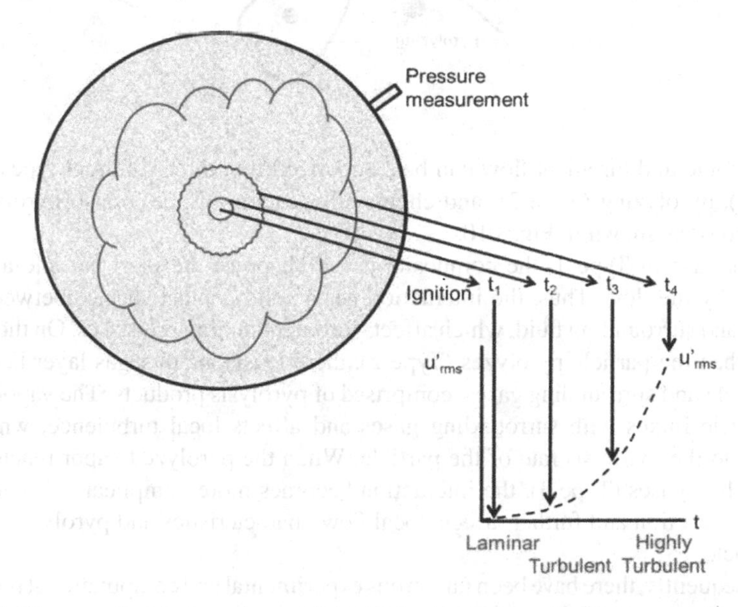

Fig. 8.9 Diagram of explosion sphere with increasing turbulence with flame propagation

the hazard is now dependent on the experimental apparatus. The problem arises mainly due to the increase in turbulent intensity caused by the expanding combustion products in a constant volume vessel [11] and the initial turbulence from the dispersal of dusts.

Figure 8.9 shows an illustrative sketch of an expanding flame front at four different times inside a typical explosion sphere [44]. The graph in Fig. 8.9 shows the turbulent intensity at different time instants. Since the flame front is non-stationary and accelerates as the flame grows in size, the turbulent intensity will also correspondingly increase as shown in Fig. 8.9.

Location t_1 indicates when the flame is initiating (usually using a chemical igniter or a spark), t_2 indicates the laminar flame propagating outward away from the ignition point, and t_3 denotes onset of turbulence which wrinkles the flame. At t_4, the turbulent intensity (u'_{rms}) further increases as shown in the inset of Fig. 8.9.

The turbulence-particle interaction is also a fascinating area of study that has not been explored much in combustion research. In general, the interactions between

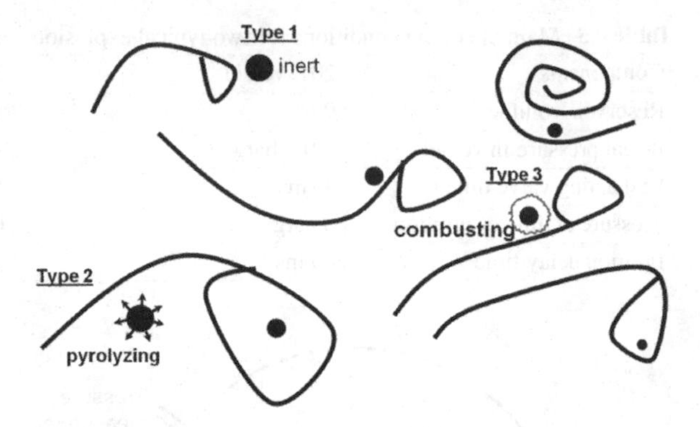

Fig. 8.10 Schematic of three different types of particle-flow interactions

solid particle and turbulent flow can be categorized into three different types, inert (Type 1), pyrolyzing (Type 2), and chemically reacting (Type 3, also pyrolyzing) interactions, as shown in Fig. 8.10.

In the case of Type 1, the composition and shape of the inert particle are not changed by the flow. Thus, the interaction has a solid contact surface between the particle and surrounding fluid, which affects turbulent mixing processes. On the other hand, when the particle pyrolyzes (Type 2), there exists another gas layer between the particle and surrounding gases, comprised of pyrolysis products. The vapor from the particle mixes with surrounding gases and affects local turbulence, which is coupled to the pyrolysis rate of the particle. When the pyrolyzed vapor reacts with surrounding gases (Type 3), the interaction becomes more complicated because of chemical reaction and further affects local flow characteristics and pyrolysis rate of the particle.

Consequently, there have been numerous experimental and computational research efforts on inert particle-flow interaction [45–49] (Type 1) due to its relative simplicity compared with other interactions (Types 2 and 3) although investigating inert particle-flow interaction is still challenging. When it comes to the pyrolyzing particle-flow interaction (Type 2), there is almost no fundamental knowledge because of the complexity of the processes.

There are no robust prediction tools developed up to date. Lacking knowledge in the pyrolyzing particle-flow interaction poses significant challenges in predicting phenomenon in other fire-dynamic problems like wildfire propagation through fire brand-related spot ignitions.

The deflagration index has also been used to estimate the laminar burning velocity for a given dust concentration and particle size using expressions such as [50] (see also Table 8.3):

$$S_L = \frac{K_{st}}{4.84\left(\dfrac{P_{max}}{P_0} - 1\right) P_{max}},$$

where 4.84 is an empirical constant, P_{max} is the maximum pressure, and P_0 is the initial pressure. Similar expressions are also used in numerical CFD codes that are used to model dust explosions [51]. Modeling codes like FLACS [52] and similar modeling programs use an empirical correlation ($S_T = f(S_L, u'_{rms}, l_0)$) [44] [60], which correlates the turbulent burning velocity (S_T) as a function of the laminar burning velocity (S_L), turbulent intensity (u'_{rms}), and integral length scale (l_0). It should be noted that none of these are measured for dust-air mixtures.

The effect of dust particles on the turbulent burning velocity has been recently studied by Rockwell [7] and Ranganathan [12]. The calculation method (per NFPA 68) for the venting of deflagration of purely gaseous (no dust) mixtures was recently revised based primarily on the fundamental burning velocity of gas mixture and can be expressed, in general, as $A_v = f(S_L, P_{max}, \lambda)$, where A_v is the minimum required venting area, S_L is laminar burning velocity of gas-air mixture, P_{max} is the maximum pressure developed in a constant volume vessel, and λ is the ratio of gas-air mixture burning velocity accounting for turbulence in vented deflagration to the fundamental laminar burning velocity. This revision was possible because of the extensive research and experimentation on laminar and turbulent burning velocity and the availability of these data for purely gaseous mixtures. However, the venting design of dust-air and hybrid mixtures still relies on K_{st} and P_{max}. These methods do not provide acceptable levels of risk in many situations due to the inconsistency in the dependent parameter values determined from different spherical vessels and they are revised constantly with the development of additional research outcomes. Thus, the future studies on burning velocity of dust-gas mixtures and hybrid mixtures are important for the future development of deflagration protection design and calculations.

8.4 Other Parameters of Relevance to the Dust Flames and Engineering Standards

Currently, NFPA 654 (Standard for Prevention of Fire and Dust Explosion from the Manufacturing, Processing and Handling of Combustible Solid Particles) and 664 (Standard for the Prevention of Fires and Explosions in Wood Processing and Woodworking Facilities) list the following parameters that are used to identify a dust explosion hazard (a description of each parameter is further elaborated in Appendix A):

1. Minimum explosion concentration, MEC.
2. Minimum ignition energy (MIE).
3. Particle size distribution.
4. Moisture content.
5. Maximum explosion pressure at optimum concentration.
6. Maximum rate of pressure rise at optimum concentration.

7. Deflagration index or K_{st}.
8. Depth of dust accumulation.
9. Layer ignition temperature.
10. Dust cloud ignition temperature.
11. Limiting oxidant concentration to prevent ignition.
12. Electrical resistivity.
13. Charge relaxation time.
14. Chargeability.
15. Minimum dust accumulation area.
16. Lowest temperature at which pyrolysis has been reported.
17. Minimum dust accumulation thickness (0.8 mm—NFPA 654 and 3.2 mm—NFPA 664).

Among the list of parameters, regulation is mostly based on the minimum dust layer thickness which is the main basis for identifying a dust hazard. The thickness characterizes the quantity of fugitive dust that can reasonably be suspended by a single credible upset *versus* the room volume it can be suspended in (see case study 1). This quantity, however, does not characterize the properties of dust and also suffers from identifying a minimum threshold. The threshold in this approach would be the quantity where the over-pressure resulting from the explosion does not result in damage beyond the initiating area. Thus, additional parameters such as size of a partial volume explosion that can be handled by the construction need to be included. Such additional parameters are included in Appendix A where a first attempt at documenting all the parameters involved in the problem (48 parameters total) is listed.

Overall, the parameters in Appendix A are grouped into thermodynamic, thermokinetic, physical, chemical, and electrical properties of the dust. In addition, facility-related parameters, such as type of construction, confinement level, operating temperature, humidity in air, etc., are also accounted. Thermodynamic properties are heat of combustion, adiabatic flame temperature, specific heat, thermal conductivity, and latent heat of vaporization. Thermokinetic properties are based on both the thermodynamic and chemical kinetic effects. For example, the laminar burning velocity depends on both thermodynamic (heat of combustion) as well as chemical kinetic (rate of gas-phase reaction) processes. Physical properties include mass of particles, particle size, ease of suspension, etc.

Chemical properties include parameters used to quantify chemical composition of the dust. Dust deflagrations are also sensitive to electrical properties of the dust which determine the ease of charge dissipation and charge storage to analyze static electric ignition. Facility-related parameters include construction type, confinement levels, etc. It could be argued that all 48 parameters listed in Appendix A are important and need to be known to quantify a hazard associated with a dust. However, a more rational engineering approach would be to make an optimal choice of these parameters based on: how important is the effect of these parameters on the explosion outcome; how variable are these parameters among most common materials susceptible to dust explosions; and how much resolution is really needed for adequate design in practice. Such questions and their answers are necessary to move dust explosion hazard analysis forward in the future.

8.5 Case Studies

Case Study 1 A 2.5 m wide by 2.5 m high 700 m long coal mine gallery has a layer of high volatility bituminous coal dust on the walls and floors of a portion of the gallery (Fig. 8.11). Assuming that the bulk density of the coal dust layer is 200 kg/m^3 and its minimum explosion concentration (MEC) (g/m^3) = 60, the dust layer thickness that would allow the MEC to be developed over the gallery cross-sectional area if all the dust were suspended as a cloud can be evaluated.

$$\text{The mass of dust in the layer} = 200\ \text{kg/m}^3 \times 3 \times 2.5 \times \delta \times L$$

$$\text{The mass of dust in the cloud} = 60 \times 10^{-3}\ \text{kg/m}^3 \times 2.5 \times 2.5 \times \delta \times L.$$

If all of the mass of dust in the layer forms the cloud to obtain an MEC in the coal gallery Eq. (1) should equal Eq. (2)

$$200 \times 3 \times 2.5 \times \delta \times L = 60 \times 10^{-3} \times 2.5 \times 2.5 \times \delta \times L \Rightarrow \delta = 0.25\ \text{mm}.$$

Thus, only a small thickness of 0.25 mm is needed to create a flammable atmosphere in the gallery. In practice, dust would also have accumulated in the ceiling in joints and trusses, and hence the thickness could be thinner. A common rule of thumb is if a one-cent coin dropped on a dust layer is not visible, then there is significant dust to pose a hazard.

Case Study 2 A laboratory tested a coal dust sample and found that a 1 g sample of coal dust comprised of carbon (solid), and methane and carbon monoxide (volatiles) given by (a) 0.42 g C, (b) 0.29 g CH_4, (c) 0.29 g CO. Using this information, the equivalence ratio of the coal dust at the MEC can be calculated. Writing down a balanced chemical equation for a stoichiometric case:

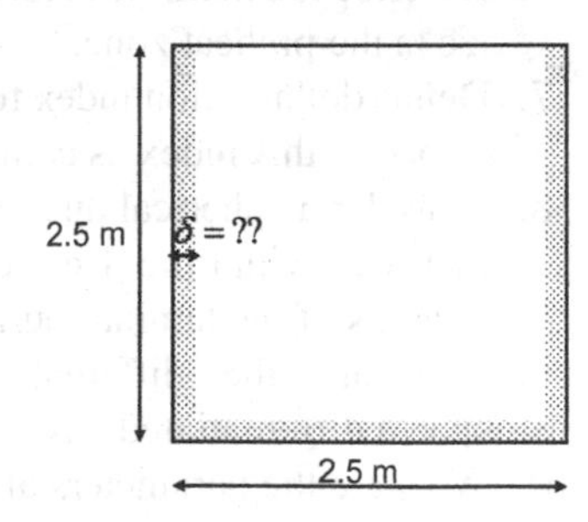

Fig. 8.11 Dust accumulation in a coal mine gallery

$$\frac{0.42\ \text{g}}{12\ \text{g/mol}}\text{C} + \frac{0.29\ \text{g}}{16\ \text{g/mol}}\text{CH}_4 + \frac{0.29\ \text{g}}{28\ \text{g/mol}}\text{CO} + x(\text{O}_2 + 3.76\text{N}_2)$$
$$\rightarrow a\text{CO}_2 + b\text{H}_2\text{O} + 3.76x\text{N}_2,$$

where x and b can be evaluated as $x = 0.0764$, $a = 0.0635$, and $b = 0.0363$ mol. Stoichiometric fuel-air ratio

$$\frac{1\ \text{gcoal dust}}{0.0764 \times 32\ g\text{O}_2 + 0.287 \times 28\ g\text{N}_2} = \frac{1}{10.5} = 0.095.$$

At a concentration of MEC (60 g/m^3), the fuel-air mass ratio is

$$\text{Fuel/air mass ratio} = \frac{60\ \text{g coal}}{(1\ m^3\ \text{air})(1200\ \text{g/m}^3)} = (20)^{-1},$$

where 1200 g/m^3 is the density of air.

$$\text{Equivalence ratio} = \frac{(\text{fuel/air})_{\text{actual}}}{(\text{fuel/air})_{\text{stoich}}} = \frac{20^{-1}}{10.5^{-1}} = 0.52.$$

Review Questions

1. What is the classification of dust particles and when is it considered hazardous?
2. Describe the controlling regions of burning for unburned dust layer in ambient air. In the case of a strong cross breeze, which parameter would be rate controlling for dust deflagration?
3. What are the flame propagation mechanisms in unburned dust-air mixtures?
4. Why does dust flame analysis require more than one equivalence ratio? Define the equivalence ratios in detail and its significance.
5. Determine the variation in the theoretical concentration for coal with ash content varying from 20% to 50%. Also evaluate and compare the theoretical concentration for coal that has 40% CH_4, 40% C, and 20% ash.
6. Describe the flame structure in dust-air flame when the particles partially vaporize in the preheat zone.
7. Define deflagration index for explosive dust clouds. What is the physical significance of this index as compared to the maximum pressure in the vessel?
8. Consider a spherical dust-air flame in an explosion vessel. The flame radius r_f increases with time. Derive an expression for velocity in the unburned mixture in terms of the laminar burning velocity, and burnt and unburnt temperature.
9. How are the different stages in flame propagation in an explosion sphere differentiated based on turbulent intensity?
10. What are the parameters affecting the turbulent burning velocity?

References

1. Sun, J., R. Dobashi, and T. Hirano. 2006. Structure of flames propagating through aluminum particles cloud and combustion process of particles. *Journal of Loss Prevention in the Process Industries* 19: 769–773.
2. Nusselt, W. 1924. *Die Verbrennung und die Vergasung der Kohle auf dem Rost* 68: 124.
3. Eckhoff, R.K. 2003. *Dust Explosions in the Process Industries*, 3rd ed. Boston: Gulf Professional Publishing.
4. Smoot, D., and M.D. Horton. 1977. Propagation of laminar pulverized coal-air flames. *Progress in Energy and Combustion Science* 3: 235–258.
5. Krazinski J.L., R.O. Buckius, and H Krier. 1979. Coal dust flames: a review and development of a model for flame propagation, *Progress in Energy and Combustion Science*, 5: 31–71.
6. Slezak, S.E., R.O. Buckius, and H. Krier. 1985. A model of flame propagation in rich mixtures of coal dust in air. *Combustion and Flame* 59: 251–265.
7. Rockwell, S.R., and A.S. Rangwala. 2013. Influence of coal dust on premixed turbulent methane-air flames. *Combustion and Flame* 160 (3): 635–640.
8. Rockwell, S.R., and A.S. Rangwala. 2013. Modeling of dust air flames. *Fire Safety Journal* 59: 22–29.
9. Xie, Y., V. Raghavan, and A.S. Rangwala. 2012. Study of interaction of entrained coal dust particles in lean methane-air premixed flames. *Combustion and Flame* 159: 2449–2456.
10. Lee, M., S. Ranganathan, and A.S. Rangwala. 2014. Influence of the reactant temperature on particle entrained laminar methane-air premixed flames. *Proceedings of the Combustion Institute* 35: 729–736.
11. Ranganathan, S., M. Lee, V.Y. Akkerman, and A.S. Rangwala. 2015. Suppression of premixed flames with inert particles. *Journal of Loss Prevention in the Process Industries* 35: 46–51.
12. Ranganathan, S., D. Petrow, S.R. Rockwell, and A.S. Rangwala. 2018. Turbulent burning velocity of methane-air-dust premixed flames. *Combustion and Flame* 188: 367–375.
13. Dahoe, A. 2000. Dust explosions: a study of flame propagation. In *Applied Sciences*, p. 298. Delft University of Technology.
14. Seshadri, K., A.L. Berlad, and V. Tangirala. 1992. The structure of premixed particle-cloud flames. *Combustion and Flame* 89 (3): 333–342.
15. Williams, F.A. 1965. *Combustion theory*. Addison-Wesley.
16. Goroshin, S., M. Kolbe, and J.H.S. Lee. 2000. Flame speed in a binary suspension of solid fuel particles. *Proceedings of the Combustion Institute* 28: 2811–2817.
17. Boichuk, L., V. Shevchuk, and A. Shvets. 2002. Flame propagation in two component aluminum-boron gas suspensions. *Combustion, Explosion, and Shock Waves* 38 (6): 651–654.
18. Ranganathan, S. 2018. Influence of dusts on premixed methane-air flames. Doctoral Thesis, Worcester Polytechnic Institute.
19. Cashdollar, K.L. 2000. Overview of dust explosibility characteristics. *Journal of Loss Prevention in the Process Industries* 13 (3–5): 183–199.
20. Ballal, D.R. 1983. Flame propagation through dust clouds of carbon, coal, aluminium and magnesium in an environment of zero gravity. *Proceedings of the Royal Society of London Series A, Mathmatical and Physical Sciences* 385: 21–51.
21. Cassel, H.M., A.K. Das Gupta, and S. Guruswamy. 1949. Factors affecting flame propagation through dust clouds. *Proceedings of the Combustion Institute* 3: 185–190.
22. Proust, C. 2006. A few fundamental aspects about ignition and flame propagation in dust clouds. *Journal of Loss Prevention in the Process Industries* 19: 104–120.
23. Dahoe, A.E., K. Hanjalic, and B. Scarlett. 2002. Determination of the laminar burning velocity and markstein length of powder-air flames. *Powder Technology* 122: 222–238.
24. Ogle, R.A., J.K. Beddow, and A.F. Vetter. 1984. A thermal theory of laminar premixed dust flame propagation. *Combustion and Flame* 58: 77–79.
25. Nomura, S.I., and T. Tanaka. 1978. Theoretical discussion of the flame propagation velocity of a dust explosion. *The Case of Uniform Dispersion of Monosized Particles Heat Transfer Japanese Research* 7: 79–86.

26. Hertzberg, M., K.L. Cashdollar, D.L. Ng, and R.S. Conti. 1982. Domains of flammability and thermal ignitability for pulverized coals and other dusts: particle size dependences and microscopic residue analyses. *Proceedings of the Combustion Institute* 19 (1): 1169–1180.
27. Hertzberg, M., I.A. Zlochower, R.S. Conti, and K.L. Cashdollar. 1987. Thermokinetic transport control and structural microscipic realities in coal and polymer pyrolysis and devolatilization: their dominant role in dust explosions. *American Chemistry Society, Division of Fuel Chemistry* 32 (3): 24–41.
28. Nagy, J., and H.C. Verakis. 1983. *Development and control of dust explosions*. New York: Marcel Dekker Inc.
29. Nagy, J., J.W. Conn, and H.C. Verakis. 1969. *Explosion development in a spherical vessel*. US Department of the Interior: US Bureau of Mines.
30. Nomura, S.I., and T. Tanaka. 1980. Prediction of maximum rate of pressure rise due to dust explosion in closed spherical and non-spherical vessels. *Industrial Engineering Chemical Process Design and Development* 19: 451–459.
31. Ogle, R.A., J.K. Beddow, and A.F. Vetter. 1983. *Numerical modelling of dust explosions: their influence of particle shape on explosion intensity in process technical program: international powder build solids handling process*. IL: Rosemont.
32. Mallard, E., and H. Le Chatelier. 1883. Thermal model for flame propagation. *Anna Mines* 4 (18): 379–568.
33. Goroshin, S., M. Bidabadi, and J.H.S. Lee. 1996. Quenching distance of laminar flame in aluminum dust clouds. *Combustion and Flame* 105: 147–160.
34. Saltelli, A., K. Chan, and E.M. Scott. 2000. *Sensitivity analysis*. Wiley, LTD.
35. ASTM-E1515. 2000. Standard test method for minimum explosible concentration of combustible dusts.
36. Andrews, G.E., D. Bradley, and S.B. Lwakabamba. 1975. Turbulence and turbulent flame propagation—a critical appraisal. *Combustion and Flame* 24: 285–304.
37. Abdel-Gayed, R.G. and D. Bradley. 1977. Dependence of turbulent burning velocity on turbulent Reynolds number and ratio of laminar burning velocity to R.M.S. turbulent velocity. *Proceedings of the Combustion Institute* 16: 1725–1735.
38. Bradley, B. 1992. How fast can we burn. *Proceedings of the Combustion Institute* 24: 247–262.
39. Eckhoff, R.K. 2005. *Explosion hazards in the process industries*. TX: Houston.
40. ISO 6184-1. 1985. Explosion Protection Systems—Part 1: determination of explosion indices of combustible dusts in air. International Organization for Standardization (ISO).
41. EN 14034-1. 2004. Determination of explosion characteristics of dust clouds - Part 1: determination of the maximum explosion pressure Pmax of dust clouds. Brussels: European Committee for Standardization (CEN).
42. EN 14034-2. 2006. Determination of explosion characteristics of dust clouds - Part 2: determination of the maximum rate of explosion pressure rise of dust clouds. Brussels: European Committee for Standardization (CEN).
43. ASTM-E1226. 2010. Standard Test Method for pressure and rate of pressure rise for combustible dusts. American Society for Testing and Materials (ASTM).
44. Rockwell, S.R., and A.S. Rangwala. 2011. Effect of coal particles on turbulent burning velocity of methane-air premixed flames. Technical Meeting of the Eastern States Section of the Combustion Institute. CT: Storrs.
45. Crowe, C.T. 2000. On models for turbulence modulation in fluid-particle flows. *International Journal of Multiphase Flow* 26: 719–727.
46. Loth, E. 2000. Numerical approaches for motion of dispersed particles, droplets and bubbles. *Progress in Energy and Combustion Science* 26 (3): 161–223.
47. Balachandar, S. and A. Prosperetti. 2007. *IUTAM symposium on computational approaches to multiphase flow: proceedings of an IUTAM symposium held at argonne national laboratory*, 4–7 Oct 2004, vol. 81. Springer Science & Business Media.
48. Prosperetti, A. and G. Tryggvason. 2009. *Computational methods for multiphase flow*. Cambridge university press.

49. Elghobashi, S. 1994. On predicting particle-laden turbulent flows. *Applied Scientific Research* 52 (4): 309–329.
50. Cashdollar, K.L. 2000. Overview of dust explosibility characteristics. *Journal of Loss Prevention in the Process Industries* 13: 183–199.
51. Skjold, T. 2007. Review of the DESC project. *Journal of the Loss Prevention in the Process Industries* 20. 291–302.
52. Arntzen, B.J., H.C. Salvesen, H.F. Nordhaug, I.E. Storvik, and O.R. Hansen. CFD modelling of oil mist and dust explosion experiments. In: *Proceedings of the 4th international seminar on fire and explosion hazards*, 601–608.

Appendix
Parameters Involved

S.No.	Parameter	Description	Established test methods or apparatus*
Thermodynamic parameters			
1	Heat of combustion (J/g)	Amount of energy released per unit mass undergoing a combustion reaction	Bomb calorimeter
2	Combustion efficiency	Fraction of energy that is utilized in pressure build up	Law of Conservation of Energy
3	Radiant heat fraction	Fraction of total heat released that is transferred via radiation mode	Radiant flux measurements
4	Latent heat of vaporization (J/g)	Amount of heat required to vaporize a unit mass of fuel	Differential Scanning Calorimeter

A. S. Rangwala and V. Raghavan, *Mechanism of Fires*,
https://doi.org/10.1007/978-3-030-75498-3

S.No.	Parameter	Description	Established test methods or apparatus*
5	Adiabatic flame temperature (°C)	Maximum possible temperature achieved by the combustion reaction in a constant pressure process	Theoretical Calculations
6	Specific heat of dust (J/g-K)	Amount of energy required per unit mass of dust to increase the temperature of the dust by one unit	Differential Scanning Calorimeter
Thermokinetic parameters			
7	Laminar burning velocity (m/s)	Velocity at which unburned gases move through a combustion front in the direction normal to the front surface	–
8	Propagation speed of smoldering reaction front (m/s)	Rate at which a exothermic oxidation reaction front moves in the direction of non-reactive zone of a dust layer	–
9	Rate of reaction in the gas phase (g/s)	Rate at which the reactant gas concentration depletes	–
10	Rate of reaction in the solid phase (surface chemical reaction rate) (g/s)	Identifies the smoldering combustion of a dust layer. Smoldering layers can release combustible vapors such as CO, CH_4, which can lead to a gas deflagration	–
11	Maximum closed volume deflagration pressure (bar)	Maximum pressure reached during a dust deflagration for the optimum concentration of the dust cloud	ASTM E1226
12	Maximum closed volume rate of pressure rise (bar/s)	Rate of pressure rise at maximum pressure reached during a dust deflagration for the optimum concentration of the dust cloud	ASTM E1226
13	Deflagration index, (K_{St}) (bar-m/s)	Rate of pressure rise at maximum pressure during a dust deflagration normalized to unit volume	ASTM E1226
14	Minimum explosion concentration (MEC) (g/m^3)	Minimum concentration of a combustible dust cloud sufficient to increase the pressure by 1 atmosphere (14.7 psi or 1.01 bar) due to deflagration. Dust assumed to be well dispersed in air	ASTM E 1515
15	Minimum ignition energy (MIE) (mJ)	Minimum energy sufficient to ignite most easily ignitable concentration of fuel in air	ASTM E 2019
16	Autoignition temperature of layer (° C)	Lowest set temperature of the surface at which dust layer on it will ignite spontaneously	ASTM E 2021

S.No.	Parameter	Description	Established test methods or apparatus*
17	Autoignition temperature of cloud (° C)	Minimum temperature at which a dust cloud will self-ignite	ASTM E 1491 06 (Godbert-Greenwald Furnace Test)
18	Limiting oxygen concentration (LOC)	Minimum oxygen concentration at the limit of flammability for the worst-case (most flammable) fuel concentration	ASTM E 2079
Physical parameters			
19	Thermal conductivity of dust (W/m° C)	Amount of heat transmitted through a unit thickness in a direction normal to a surface of unit area caused due to a unit temperature gradient	
20	Mass of combustible particulate solid (g)	Typically a fugitive dust layer can contain inerts which are not combustible	
21	Particle shape	Quantitatively, shape factors, and coefficients are used as parameters in equations governed by particle shape	Pattern recognition techniques
22	Particle size (m)	Characteristic dimension of irregularly shaped particle representing the diameter of equivalent sphere	Image Analysis with Microscope
23	Particle size distribution	Statistical term that quantifies fluctuations in size and shape of particles of given dust sample	ASTM B761-06
24	Bulk density (g/cm^3)	Weight of dust per unit volume	
25	Porosity	Measure of difference in densities of dust bulk and dust particle because of void spaces between particles in the bulk	
26	Degree of compaction of powder	Ratio of volume under specified pressure to volume under ambient pressure for a given mass of dust and configuration of dust pile	
27	Moisture content in dust	Weight percentage of water content in given dust sample	
28	Layer thickness (mm)	Minimum thickness of dust layer of a give particle size needed to cause a deflagration	
29	Surface area/volume ratio of dust (1/m)	Ratio of surface area to volume of given dust particles can be used to relate the arbitrary particle shapes to standard shapes like cube, sphere, cylinder, etc.	

S.No.	Parameter	Description	Established test methods or apparatus*
30	Suspension	Ease with which particles can be suspended in air	
31	Dispersibility	Degree of dispersion in a dust cloud, depends on cohesiveness of particles, settling velocity, moisture content	ASTM E 1945
32	Agglomeration	A mass conserving, number-reducing process that shifts the particle size distribution towards larger sizes	
33	Terminal settling velocity of dust particle (m/s)	Velocity of a particle when the drag force and buoyancy force balance equal the gravitational pull	
34	Speed of sound in dust cloud (m/s)	Plays an important role in all compressible flow phenomena	
Chemical parameters			
35	Chemical composition	Molecular formula of the sample gives important information like molecular weight, acidic or basic nature, special affinity for other chemicals	
36	Reactivity with water		
Electrical parameters			
37	Volume resistivity	Measure of electrostatic ignition hazard of the dust	IEC 60093
38	Charge relaxation time	Time duration of charge retention in a dust	IEC 61340-2-1:2000
39	Chargeability	Propensity of dust particles to become charged when flowing or air borne	IEC 61340-2-1:2000
External parameters (facility related)			
40	Size of partial volume explosion that can be handled by the construction	This factor will depend on construction type, volume of initial cloud that can be formed, number of vents installed, and nature of dust	
41	Type of construction	Based on NFPA 220 standard on types of building construction	
42	Room volume (m^3)	Total volume of room/enclosure where fugitive dust accumulation is possible	
43	Operating temperature (°C)	Certain facilities could operate at a temperature higher than ambient. Thus possibility of autoignition is higher.	
44	Operating pressure (bar)	Certain facilities can operate at pressures other than atmospheric. Studies have shown that thermodynamic and thermokinetic properties vary with pressure	

S.No.	Parameter	Description	Established test **methods or** apparatus*
45	Relative humidity	Major of quantity of water vapor in ambient air	
46	Confinement	Dimensions of the enclosure which is considered to be at constant temperature and pressure and surrounds given test apparatus or control volume under consideration	
47	Turbulence	Flow instability represented by chaotic state of fluid motion with dissipative structure	
48	Detonability limit	Condition outside which self-sustained propagation of detonation wave cannot be realized	

* Test methods starting with ASTM and IEC are standard test methods (some standard test methods are not designed for dust per se but can be easily modified to include dust samples)

i. ASTM B761-06: Standard Test Method for Particle Size Distribution of Metal Powders and Related Compounds by X-ray Monitoring of Gravity Sedimentation

ii. ASTM E 1226-05: Standard Test Method for Pressure and Rate of Pressure Rise for Combustible Dust

iii. ASTM E 1491-06: Standard Test Method for Minimum Autoignition Temperature of Dust Clouds

iv. ASTM 1515-07: Standard Test Method for Minimum Explosible Concentration of Combustible Dusts

v. ASTM E 1945-02 (2008): Standard test Method for Percent Dispersibility

vi. ASTM E 2019-03 (2007): Standard Test Method for Minimum Ignition Energy of a Dust Cloud in Air

vii. ASTM E 2021-06: Standard Test Method for Hot-Surface Ignition Temperature of Dust Layers

viii. ASTM E 2079-07: Standard Test Methods for Limiting Oxygen (oxidant) Concentration in Gases and Vapors

ix. IEC 60093: Methods of test for volume resistivity and surface resistivity of solid electrical insulating materials

x. IEC 61340-2-1 (2002-06): Measurement methods—Ability of materials and products to dissipate static electric charge

xi. IEC 61340-2-2 (2000-067): Measurement methods—Measurement of chargeability

Index

A. S. Rangwala and V. Raghavan, *Mechanism of Fires*,
https://doi.org/10.1007/978-3-030-75498-3

Printed in the United States
by Baker & Taylor Publisher Services